Natural Antioxidants: Determination in Food and Nutraceuticals and Implications on Human Health

Natural Antioxidants: Determination in Food and Nutraceuticals and Implications on Human Health

Guest Editor

Gregorio Peron

MDPI

Basel • Beijing • Wuhan • Barcelona • Belgrade • Novi Sad • Cluj • Manchester

Guest Editor
Gregorio Peron
Department of Molecular and
Translational Medicine
University of Brescia
Brescia
Italy

Editorial Office
MDPI AG
Grosspeteranlage 5
4052 Basel, Switzerland

This is a reprint of the Special Issue, published open access by the journal *BioTech* (ISSN 2673-6284), freely accessible at: https://www.mdpi.com/journal/biotech/special_issues/7KP0712M2L.

For citation purposes, cite each article independently as indicated on the article page online and as indicated below:

Lastname, A.A.; Lastname, B.B. Article Title. *Journal Name* **Year**, *Volume Number*, Page Range.

ISBN 978-3-7258-8061-4 (Hbk)
ISBN 978-3-7258-8062-1 (PDF)
https://doi.org/10.3390/books978-3-7258-8062-1

Contents

About the Editor

Gregorio Peron

Gregorio Peron is Associate Professor of Food Chemistry at the Department of Molecular and Translational Medicine of the University of Brescia, Italy. He obtained his Master's Degree in Medicinal Chemistry from the University of Padova in 2014 and completed his Ph.D. in Pharmaceutical Sciences at the same institution in 2018. His research activity focuses on metabolomics, phytochemistry, nutraceuticals, and the characterization of natural products and food systems through advanced analytical techniques, including chromatography and mass spectrometry. Prof. Peron is particularly interested in the identification of bioactive compounds from medicinal and edible plants, antioxidant evaluation, and the valorization of agri-food byproducts through sustainable and innovative approaches such as microbial fermentation. He is also involved in interdisciplinary research on food quality, functional foods, and the biological effects of natural compounds related to human health.

BioTech

Editorial

Natural Antioxidants: Determination in Food and Nutraceuticals and Implications on Human Health

Gregorio Peron

Department of Molecular and Translational Medicine, University of Brescia, 25123 Brescia, Italy; gregorio.peron@unibs.it

Oxidative processes influence several aspects of biology, from the subtle balance of redox signaling to the destructive cascade of oxidative damage associated with chronic disease and aging. In this context, natural antioxidants, i.e., bioactive compounds derived from food matrices and nutraceutical sources, have attracted scientific and public interest for their potential in promoting health, preserving food quality, and mitigating adverse outcomes associated with oxidative stress [1]. This Special Issue (SI), titled "Natural Antioxidants: Determination in Food and Nutraceuticals and Implications for Human Health", was conceived to address both foundational and translational aspects of this rapidly evolving field. The significance of natural antioxidants lies in their multifaceted roles. Chemically, they are capable of neutralizing reactive molecular species, thereby reducing direct oxidative damage to lipids, proteins, and nucleic acids. Biologically, they influence endogenous defense systems through modulation of redox-sensitive signaling pathways and detoxifying enzymes [2]. Furthermore, the metabolic interplay between dietary antioxidants and the gut microbiota suggests complex systemic effects that extend beyond direct free-radical scavenging [3]. However, despite widespread interest, their mechanisms of action and physiological relevance remain areas of active investigation and debate.

A fundamental aspect of antioxidant research relies on the accurate determination and characterization of these compounds in food and nutraceuticals. Reliable analytical strategies (ranging from chromatographic separation to advanced spectrometric profiling) are required for defining antioxidant composition, quantifying bioactive components, and assessing their stability during food processing, storage, and digestion. This SI has collected contributions that reinforce the critical link between methodological rigor and meaningful biological interpretation. Across diverse food matrices and biological models, the findings presented in the SI highlight the diversity of antioxidant phenotypes, but also the necessity of integrative analytical frameworks to reveal their functional potential.

From a technological perspective, insights into how processing and preparation influence antioxidant profiles were emphasized. Natural products often undergo modifications during plant germination [4], extraction procedures, or enzymatic treatment [5], which can enhance or impair antioxidant capacity. Understanding the effects of such manipulations is crucial, both for the optimization of functional food ingredients and for preserving the intrinsic nutritional value of traditional dietary sources. These considerations extend to the formulation of nutraceuticals where extraction, fractionation, and delivery systems determine their bioavailability and biological efficacy in vivo.

Interdisciplinary approaches for studying antioxidants are presented in this SI, ranging from phytochemical quantification to microbial metabolite profiling and in vivo bioactivity assessments. The collective work clearly demonstrates that effective antioxidant research

 https://doi.org/10.3390/biotech15010010

should integrate diverse yet complementary disciplines such as analytical chemistry, food science, microbiology, and biological evaluation. This integrative perspective is appropriate because antioxidant function cannot be fully understood solely through chemical assays. Indeed, it requires contextualization within complex biological systems, including cellular models, animal models, and whole-organism responses.

The human health implications of natural antioxidants remain a central focus. Oxidative stress is involved in the pathogenesis of cardiovascular diseases, metabolic disorders, neurodegeneration, and cancer [2]. Natural antioxidants, whether delivered through diet or nutraceutical supplementation, are often promoted for their purported preventive or therapeutic benefits. However, elucidating how these compounds behave physiologically remains an ongoing challenge. The contributions in this SI explore aspects of this continuum, including the dynamics of antioxidant uptake and clearance, and modulation of enzymatic defenses in living organisms, thus providing valuable data to inform hypotheses on their health relevance.

Another aspect of antioxidant research covered in this SI is the relationship between natural antioxidants and the gut microbiome. The transformation of dietary antioxidants by intestinal microbes can yield metabolites with distinct bioactivities, suggesting that the health implications of antioxidant intake cannot be fully appreciated without considering the role of microbial metabolism. This emerging area bridges nutritional biochemistry and microbiome science, pointing toward more nuanced conceptual models of how dietary components influence systemic physiology.

In addition to their health benefits, natural antioxidants play important roles in food quality and preservation. Oxidative degradation can compromise sensory attributes, nutritional value, and shelf life. Antioxidants can contribute to food stability and consumer appeal by mitigating lipid peroxidation and preserving functional compounds. This dual relevance to food science and health highlights the interconnected nature of research in this domain.

Taken together, the contributions to this SI reflect a comprehensive exploration of antioxidants from chemical determination to biological implication. They demonstrate that progress in this field depends on methodological excellence in analytical determination, conceptual clarity about biological mechanisms, and translational insight into human health implications. By emphasizing diverse food sources and analytical strategies, and integrative biological approaches, this collection advances our understanding of how natural antioxidants can be characterized, optimized, and potentially leveraged for health promotion. As the scientific community continues to refine analytical tools and deepen biological models, future research will further clarify the conditions under which natural antioxidants exert health effects in vivo, how they interact with complex biological systems, and how food science innovations can enhance their functional potential.

Finally, as the Guest Editor, I would like to thank all the authors who contributed to this SI by publishing their research, as well as the reviewers and the Assistant and Academic Editors for their valuable support.

Conflicts of Interest: The author declares no conflict of interest.

References

1. Gulcin, İ. Antioxidants: A comprehensive review. *Arch Toxicol.* **2025**, *99*, 1893–1997. [CrossRef] [PubMed]
2. Chandimali, N.; Bak, S.G.; Park, E.H.; Lim, H.-J.; Won, Y.-S.; Kim, E.-K.; Park, S.-I.; Lee, S.J. Free radicals and their impact on health and antioxidant defenses: A review. *Cell Death Discov.* **2025**, *11*, 19. [CrossRef] [PubMed]
3. Munteanu, C.; Schwartz, B. Interactions between Dietary Antioxidants, Dietary Fiber and the Gut Microbiome: Their Putative Role in Inflammation and Cancer. *Int. J. Mol. Sci.* **2024**, *25*, 8250. [CrossRef] [PubMed]

https://doi.org/10.3390/biotech15010010

4. Günal-Köroğlu, D.; Esatbeyoglu, T.; Capanoglu, E. Effect of germination on the phenolic compounds: Content, bioavailability, food applications, and health benefits. *Food Meas.* **2025**, *19*, 8144–8164. [CrossRef]
5. Peron, G.; Ferrarese, I.; Carmo Dos Santos, N.; Rizzo, F.; Gargari, G.; Bertoli, N.; Gobbi, E.; Perosa, A.; Selva, M.; Dall'Acqua, S. Sustainable Extraction of Bioactive Compounds and Nutrients from Agri-Food Wastes: Potential Reutilization of Berry, Honey, and Chicory Byproducts. *Appl. Sci.* **2024**, *14*, 10785. [CrossRef]

https://doi.org/10.3390/biotech15010010

BioTech MDPI

Article

Metabolites Potentially Determine the High Antioxidant Properties of *Limosilactobacillus fermentum* U-21

Yelena V. Grishina [1,2], Aleksey A. Vatlin [1], Dilara A. Mavletova [1], Maya V. Odorskaya [1], Alexey M. Senkovenko [3], Rustem A. Ilyasov [1,4,*] and Valeriy N. Danilenko [1]

1 Laboratory of Bacterial Genetics, Vavilov Institute of General Genetics, Russian Academy of Sciences, 119333 Moscow, Russia; grishina.e@phystech.edu (Y.V.G.); vatlin_alexey123@mail.ru (A.A.V.); lmavletova@mail.ru (D.A.M.); maya_epifanova@mail.ru (M.V.O.); valerid@vigg.ru (V.N.D.)
2 Moscow Institute of Physics and Technology, National Research University, 141701 Moscow, Russia
3 Department of Bioengineering, Faculty of Biology, M.V. Lomonosov Moscow State University, Leninskie Gory 1/12, 111234 Moscow, Russia; senkoven99@gmail.com
4 Laboratory of Molecular Genetics, Bashkir State Agrarian University, 450001 Ufa, Russia
* Correspondence: apismell@hotmail.com

Abstract: Many kinds of *Lactobacillus* are common occupants of humans' digestive tract that support the preservation of a balanced microbial environment that benefits host health. In this study, the unique lactic acid bacterium strain *Limosilactobacillus fermentum* U-21, which was isolated from the feces of a healthy human, was examined for its metabolite profile in order to compare it to that of the strain *L. fermentum* 279, which does not have antioxidant (AO) capabilities. By using GC × GC−MS, the metabolite fingerprint of each strain was identified, and the data were then subjected to multivariate bioinformatics analysis. The *L. fermentum* U-21 strain has previously been shown to possess distinctive antioxidant properties in in vivo and in vitro studies, positioning it as a drug candidate for the treatment of Parkinsonism. The production of multiple distinct compounds is shown by the metabolite analysis, demonstrating the unique characteristics of the *L. fermentum* U-21 strain. According to reports, some of the *L. fermentum* U-21 metabolites found in this study have health-promoting properties. The GC × GC−MS-based metabolomic tests defined strain *L. fermentum* U-21 as a potential postbiotic with significant antioxidant potential.

Keywords: *Limosilactobacillus*; antioxidant; immunomodulatory activity; gut microbiota; GC-MS; metabolomics

Key Contribution: The lactic acid bacteria strain *Limosilactobacillus fermentum* U-21 metabolites showed unique antioxidant activity.

1. Introduction

Advanced multiomics technologies, which are largely focused on the universal detection of genes (genomics), mRNA (transcriptomics), proteins (proteomics), and metabolites (metabolomics) in a biological sample, have helped enhance studies on the gut microbiota. Multiomics technology enables researchers to study the gut microbiome with a better understanding of how the host and microbial community interact. The most recent and quickly developed branch of omics is metabolomics, which provides information about the biological and metabolic processes of microorganisms [1]. Metabolomics technologies allow studying molecular mechanisms of host-bacterial interactions and the mechanisms by which the hosts of the microbiota affect the host's immunological status [2].

Researchers' interest in *Lactobacillus* species' metabolomics analysis has grown because these bacteria are the most common occupants of the gastrointestinal tracts of animals, including humans [3–5]. Lactobacillus is a species in the gut microbiome that is a natural source of postbiotics because of the active synthesis of biologically active compounds and the bidirectional communication with the host organism [6–10]. As a result, *Lactobacillus* can affect the immune status and antioxidant (AO) status of the host. Postbiotics are metabolites or components produced by microbiota that have an impact on human health [11]. Antioxidant properties of many *Lactobacillus* species have already been confirmed by many in vivo and in vitro studies [12–15]. The *Limosilactobacillus fermentum* U-21 strain used in this study is a potential pharmabiotic. The strains were selected previously based on their ability to reduce oxidative stress in vitro on a bioluminescent test system based on Escherichia coli MG1655 strains carrying plasmids encoding luminescent biosensors pSoxS-lux and pKatG-lux inducible by superoxide anion and hydrogen peroxide [13]. The antioxidant properties of the strain were next confirmed using in vivo and in vitro methods. In particular, *L. fermentum* U-21 prolonged the life by 25% of the nematode *C. elegans* under paraquat-induced oxidative stress [14]. *L. fermentum* U-21 stopped the degenerative alterations in internal organs [8] and the degeneration of brain dopaminergic neurons in a mouse model of paraquat-induced Parkinsonism [14]. The twenty-nine genes related to the antioxidant properties, which are involved in the thioredoxin complex, as well as heavy metal metabolism and transport, were found in the genome of the *L. fermentum* U-21 strain [16]. Comparative analysis of the proteins in the culture fluid of the *L. fermentum* U-21 strain and *L. fermentum* 279 strain, which does not exhibit AO properties [13,14], revealed significant differences associated with LysM and ATP-binding Clp protease [16]. ATP-binding Clp protease may be essential for the refolding of misfolded proteins caused by oxidative stress [17,18]. Both of the L. fermentum strains under study have the LysM protein, which can bind to peptidoglycan and affect the generation of humoral immune components [19]. The *L. fermentum* U-21 strain, however, has the short serine-threonine insertion [16]. The gathered data have shown that *L. fermentum* U-21 has special antioxidant characteristics, making it a promising candidate as a postbiotic [8,9,12].

In this study, we examine the metabolomic profile of the *L. fermentum* U-21 strain—candidate for a postbiotic—using an untargeted metabolomic technique. In addition, we compare the metabolites and related metabolic pathways among the *L. fermentum* U-21 strain, which demonstrates high AO properties, and the *L. fermentum* 279 strain, which does not exhibit AO properties. The main purpose of this study was to identify the unique metabolites of *L. fermentum* U-21 compared to *L. fermentum* 279, which does not exhibit AO properties.

2. Materials and Methods

2.1. Bacterial Strains

We used *L. fermentum* U-21 and *L. fermentum* 279 in this investigation, both of which were isolated from the bodies of people who lived in the Central European region of the Russian Federation. Both strains were preserved at the collection of the laboratory of genetics of microorganisms, Vavilov Institute of General Genetics Russian Academy of Sciences. The *L. fermentum* U-21 and 279 strains were characterized (Table 1).

Table 1. Genetic characterization of strains.

Strain	BioSample ID	GenBank Sequence
Limosilactobacillus fermentum U-21	SAMN08290293	GCA_002869825.2
Limosilactobacillus fermentum 279	SAMN08014151	GCA_002794275.1

2.2. Method, Conditions, and Media for Strain Propagation

The *L. fermentum* U-21 and *L. fermentum* 279 strains were both grown on the MRS media (Himedia) under partially anaerobic conditions (in a desiccator where oxygen was burned up by burning a candle). The MRS media contained 10.0 g/L proteosopeptone, 10.0 g/L meat extract, 5.0 g/L yeast extract, 20.0 g/L glucose, 1.0 g/L polysorbate 80, 2.0 g/L ammonium citrate, 5.0 g/L sodium acetate, 0.1 g/L magnesium sulfate, 0.05 g/L manganese sulfate, and 2.0 g/L dibasic potassium phosphate (pH 6.5 at 25 °C). For metabolomic analysis, the bacterial cultures were grown to the stationary growth phase (OD600 = 2.5). The cultivation temperature was 37 °C.

2.3. Phase Separation

Culture fluid of *L. fermentum* U-21 and *L. fermentum* 279 were prepared using the methanol-chloroform method with some modifications. All stages of selection pass at 4 °C. Bacterial cells were separated from the culture liquid at 7000× *g* for 10 min at 4 °C. After centrifugation, the culture liquids were filtered through a PES membrane (0.22 μm). The 40 mL of the supernatant was concentrated using Amicon Ultra-4 5000 MWCO. For further phase separation we selected 200 μL. Phase of culture fluid was separated by addition of 1 volume of methanol (*v*/*v*, 20 °C) and 2 volumes of chloroform, vortexed, and incubated for 20 min on ice. After incubation, we added 0.75 volume of Milli-Q water and incubated for 10 min on ice. After centrifugation at 8000× *g* for 10 min at 4 °C for phase separation, the aqueous phase was transferred to a 1.5 mL microcentrifuge tube. The methanol-chloroform method was conducted again for the upper lipid-containing phase. All samples, including extracellular vesicles, the aqueous, and lipid-containing phases, were evaporated using a SpeedVac concentrator.

2.4. Extracellular Vesicles Preparation and Physicochemical Analysis

Bacterial cells were separated from the culture liquid at 7000× *g* for 10 min at 4 °C. After centrifugation, the culture liquids were filtered through a PES membrane (0.22 μm). The 60 mL of supernatant was used to isolate the extracellular vesicles using ultracentrifugation at 260,000× *g* for 1 h 40 min at 4 °C and resuspended in phosphate-buffered saline (PBS, pH 7.4). Using a Qubit 3.0 fluorometer, the DNA, RNA, and protein quantities in the extracellular vesicles were quantified (Life Technologies, Grand Island, NY, USA). Extracellular vesicles in PBS (pH 7.4) suspension were deposited on a carbon film-coated copper grid treated by glow discharge and then stained with 1% uranyl for 3 min in an acetate solution for negative staining. Using the JEM-1400 (JEOL Inc., Tokyo, Japan) transmission electron microscope, the prepared grid was examined.

2.5. Extraction of Metabolites

Further sample preparation was carried out according to the protocol of O. Fiehn [20]. Then the metabolites were extracted with a mixture of isopropanol, acetonitrile, and water (3:3:2, *v*/*v*/*v*). An aliquot was evaporated to dryness, and then redissolved in a mixture of acetonitrile and water (1:1, *v*/*v*); the supernatant was evaporated to dryness.

2.6. Derivatization of Metabolites

In total, 10 μL of 20 mg/mL methoxyamine pyridine hydrochloride was added for derivatization, and the mixture was then shaken vigorously for 1.5 h at 30 °C. The samples were then further derivatized by adding 91 μL of a mixture of MSTFA and FAME to each sample, which was then incubated at 37 °C for 30 min on a thermal shaker. For $GC \times GC - MS$ analysis, the derivatized samples were put in a sample bottle.

2.7. GC × GC-MS Analysis

The GC-MS system consisted of an Agilent 7890/M780EI gas chromatograph (GC, Agilent Technologies, Santa Clara, CA, USA) coupled with a PERSEE mass spectrometer (MS, LECO, St. Joseph, MI, USA) and an L-PAL3 autosampler (PAL Systems, Zwingen, Switzerland). A sample of 1.0 μL was injected into a Restek Rxi-5Sil MS capillary column (30 m in length, Restek, Bellefonte, PA, USA) with a split ratio of 100:1 or 50:1. The column of the second dimension—Restek Rxi-17Sil MS (Restek, Bellefonte, PA, USA)—was 3 m 1in length. After sample preparation, a 70-eV electron beam was used to heat the electron impact (EI) ion source to 250 °C. Then, the solvent delay was 350 s, and the mass spectrometer transfer line temperature was 280 °C. The flow rate of the helium carrier gas was fixed to 1.0 mL/min. Following a 1 min hold at 60 °C, the GC oven ramped up to 280 °C at a rate of 10 °C/min. With a speed of 200 specters per second, masses were obtained in full scan mode spanning the range of 35 to 700 m/z. The chromato-mass spectrometer was controlled using the ChromaTOF software (v. 5.51, LECO, St. Joseph, MI, USA).

2.8. Data analysis

The resulting spectrum files were processed in ChromaTOF (v. 5.51, LECO, St. Joseph, MI, USA) for deconvolution, peak selection, alignment, and search in the primary database. Metabolites were identified based on mass spectra and retention times from the National Institute of Standards and Technology (NIST) libraries, the Mainlib and Feign libraries, and the National Institutes of Health (NIH) public repository.

We used ChromaTOF Tile v.1.01 (LECO, St. Joseph, MI, USA) to reduce the multidimensionality of experimental data based on the Fisher coefficient and identify significantly different chemicals in the culture liquids. The processing principle of ChromaTOF Tile v.1.01 is the comparison of two matching sections of the chromatogram (so-called tiles) and highlighting the low and high levels. The size of the studied cells was 3 × 24 in modulation and spectral measurements, respectively. Only results with a signal-to-noise ratio greater than 70 were counted. The range of analyzed masses was limited from m/z = 85 to m/z = 700. Identification was performed using the NIST database for mass spectra and retention indices (mainlib, replib) and the Leco-Fiehn rtx5 library. The matches with a direct and reverse similarity of more than 700 were considered significant.

Values with an F-ratio (intra-/inter-class variation ratio) greater than 5 were counted. A high F-ratio is predicted for a metabolite that varies between classes but is consistent within a class (low inter-class variation). The samples with a high F-ratio will be assigned to a different class (with three technical repeats and two controls of blank and pure methanol). The data were processed using the Python package, Pandas. N-Trimethylsilyl-N-Methyl Trifluoroacetamide (MSTFA), as well as its derivatives, were not taken into account. Duplicated metabolites were not counted. The samples were classified by the names of the metabolites. The metabolites were sorted by strains and fractions. Metabolite Set Enrichment Analysis (MSEA) was performed to distribute metabolites along metabolic pathways using the Metaboanalyst (https://www.metaboanalyst.ca/, accessed on 10 January 2023).

3. Results

3.1. Characteristic of the Metabolite Profile of the Culture Supernatant

Untargeted metabolite profiles of the samples from the *L. fermentum* U-21 strain and the strain *L. fermentum* 279, which does not exhibit AO properties, identified 681 metabolites in the aqueous phase of *L. fermentum* U-21 and 98 in the lipid-containing phase. Altogether, 257 metabolites of the total metabolite amount in the aqueous phase and 60 metabolites in the lipid-containing phase appeared to be the unique metabolites of *L. fermentum* U-21 in comparison with *L. fermentum* 279 (Figure 1).

The main classes of compounds detected included primary sugars, organic acids, amino acids in the aqueous phase and mostly fatty acids, and lipids in the lipid-containing phase. All common amino acids, except arginine, were detected in the aqueous phase of *L. fermentum* U-21; in the lipid-containing phase, alanine, asparagine, threonine, proline, and tryptophan were detected. There were nineteen organic acids in the aqueous phase and eight in the lipid-containing phase.

Twenty-four metabolites were reported in the literature to possess health-improving properties, of which six metabolites were amino acids. The remaining 18 metabolites were vitamin B3, vitamin B5, vitamin B6, L-Dopa, dopamine, noradrenaline, adrenaline, and GHB, which belong to different classes such as vitamins and neurotransmitters. Full, detailed results for the detected metabolites are shown in Tables 2–4 below and in Supplementary Tables S1–S4.

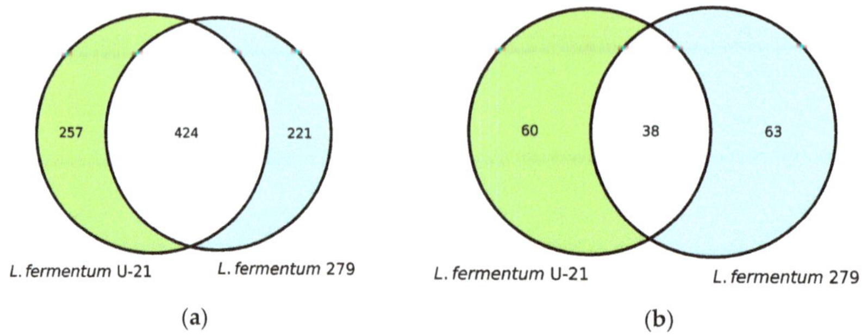

Figure 1. Venn diagrams show shared and unique metabolite features and their numbers for: (**a**) the aqueous phase of *L. fermentum* U-21 and *L. fermentum* 279; (**b**) the lipid-containing phase of *L. fermentum* U-21 and *L. fermentum* 279.

Table 2. Main classes of metabolites detected in the aqueous phase of culture supernatant of *L. fermentum* U-21.

Metabolite	Class
Glycine (Gly)	Amino acids and derivatives
Alanine (Ala)	
Valine (Val)	
Leucine (Leu)	
Isoleucine (Ile)	
Aspartic acid (Asp)	
Asparagine (Asn)	
Glutamic acid (Glu)	
Glutamine (Gln)	
Serine (Ser)	
Threonine (Thr)	
Methionine (Met)	
Cysteine (Cys)	
Lysine (Lys)	
Histidine (His)	
Proline (Pro)	
Phenylalanine (Phe)	
Tyrosine (Tyr)	
Tryptophan (Trp)	
2-Aminocaprylic acid	
Norleucine	
Homocysteine	
Homoserine	
Norvaline	
Ornithine	
N-Methyl-α-aminoisobutyric acid	
β-Alanine	
5-Hydroxytryptophan	
Pyroglutamic acid	
Aceturic acid	
α-Methyl-L-tyrosine	
N,*N*-Dimethylglycine	
N-Acetyl-L-glutamic acid	
N-Acetylglutamine	
N-Acetyl-L-phenylalanine	
N-Acetyltyrosine	

Table 2. *Cont.*

Metabolite	Class
Carnitine Propanoic acid Cinnamic acid Succinic acid Formic acid Fumaric acid Maleic acid Malonic acid Phthalic acid Propanedioic acid Oxalic acid 2-Aminobutanoic acid 2-Hydroxybutyric acid 3-Hydroxybutyric acid Glycolic acid Lactic Acid Tartaric acid Methylcitric acid Acetic acid Hydracrylic acid Dodecanedioic acid	Organic acid
Maltose Melibiose Mannobiose Cellobiose Lactose Turanose Xylofuranose 2-Amino-2-deoxyhexose Talose Erythrose Xylose Galactose Glucose Mannose Lyxose Ribose Threose Tagatose Fructose Psicose 2-Keto-gluconic acid Gluconic acid Mannonic acid Ribonic acid Arabinonic acid Galacturonic acid Gulonic acid Xylonic acid D-Erythro-Pentitol Mannitol Glycerol meso-Erythritol 1-Deoxypentitol Threitol	Saccharides

Table 2. *Cont.*

Metabolite	Class
Pentitol Arabinitol Erythritol Arabitol Arabinofuranose Dihydroxyacetone Levoglucosan Myo-Inositol Trehalose-6-phosphate Mannopyranose Glucose oxime Mannose oxime Arabinopyranose	
2-Hydroxyisocaproic acid 4-Hydroxybutanoic acid Nonanoic acid Butanoic acid	Fatty acids and analogues

Underlined metabolites are unique for *L. fermentum* U-21.

Table 3. Main classes of metabolites detected in the lipid-containing phase of culture supernatant of *L. fermentum* U-21.

Metabolite	Class
Alanine (Ala) Asparagine (Asn) Threonine (Thr) Proline (Pro) Tryptophan (Trp) L-Norvaline 1-Aminocyclopentanecarboxylic Pyroglutamic acid 2-Hydroxy-3-methylbutyric acid 2-Methylalanine *N*-Methyl-L-glutamic acid	Amino acids and derivatives
Oxalic acid Phthalic acid Propanedioic acid Succinic acid Ritalinic acid 3-Hydroxyphenylacetic acid Acetic acid 4-Hydroxybenzoic acid	Organic acids
Lyxose Fructopyranose Mannitol meso-Erythritol	Saccharides
Heptadecanoic acid Undecanoic acid Butanoic acid 2-Hydroxyisocaproic acid Oleic acid	Fatty acids and analogues

Underlined metabolites are unique for *L. fermentum* U-21.

Table 4. Health beneficial compounds detected in the culture supernatant of *L. fermentum* U-21.

Metabolite	Class
Dopamine Histamine Noradrenaline Adrenaline GABA L-DOPA Glycine 4-Hydroxybutanoic acid (GHB)	Neurotransmitter
Indole-3-acetamide Indole Indolelactate Indole-3-ethanol	Indoles and derivatives
Tartaric acid β-Alanine Carnitine	Antioxidant
Cinnamic acid L-Norvaline	Anti-inflammatory
5-Hydroxytryptophan	Antidepressant
Pyridoxine	Vitamin B6
Niacin Niacinamide	Vitamin B3
Pantothenic acid	Vitamin B5
Acetamide	Antimicrobial

Underlined metabolites are unique for *L. fermentum* U-21.

The metabolites were separated by metabolic pathways using the Metabolite Set Enrichment Analysis (MSEA). Figures 2 and 3 show the MetaboAnalyst 5.0 pathway enrichment analysis used to evaluate the metabolic pathways most implicated in the results.

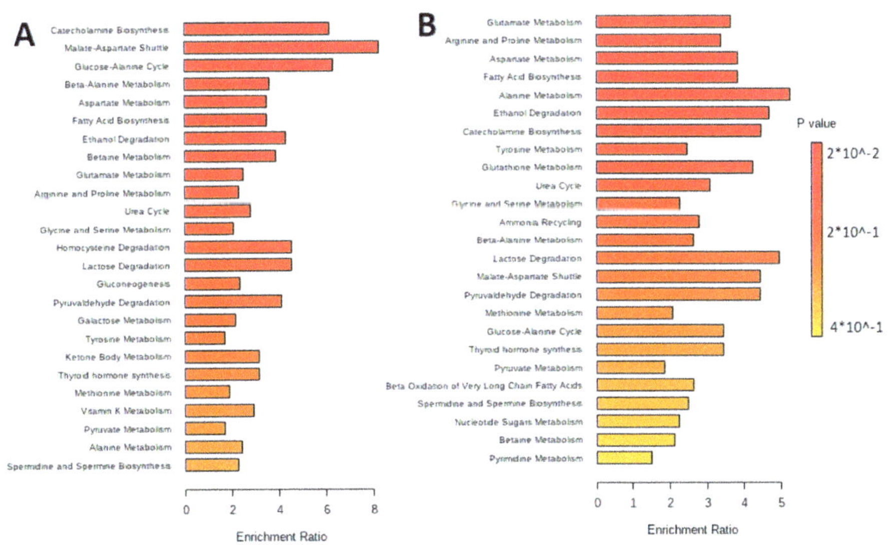

Figure 2. Pathway enrichment analysis for metabolites differed significantly between groups according to MetaboAnalyst 5.0 ($p < 0.05$). The number of hits inside a specific metabolic pathway divided by the anticipated number of hits yields the enrichment ratio. (**A**) The aqueous phase of *L. fermentum* U-21 culture supernatant. (**B**) The aqueous phase of *L. fermentum* 279′s culture supernatant.

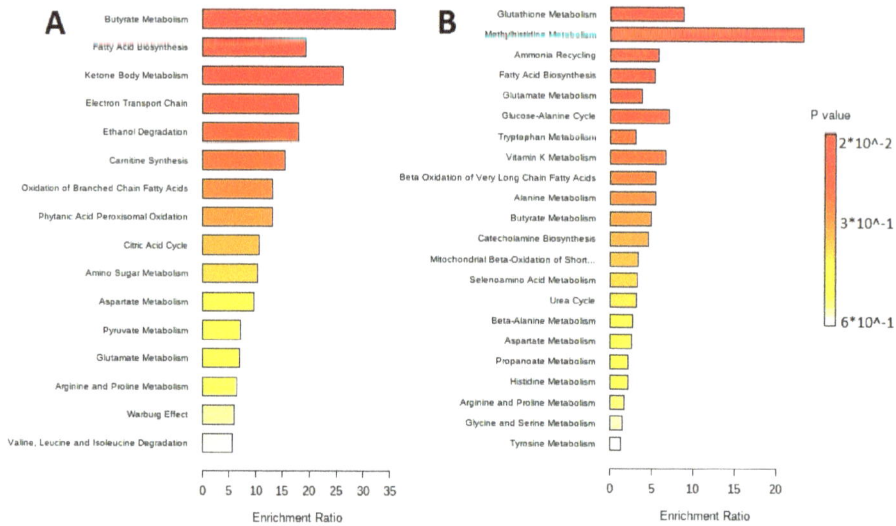

Figure 3. Pathway enrichment analysis for metabolites differed significantly between groups according to MetaboAnalyst 5.0 ($p < 0.05$). The number of hits inside a specific metabolic pathway divided by the anticipated number of hits yields the enrichment ratio. (**A**) The lipid-containing phase of *L. fermentum* U-21 culture supernatant. (**B**) The lipid-containing phase of *L. fermentum* 279′s culture supernatant.

3.2. Physicochemical Characterization of *L. fermentum* U-21 Extracellular Vesicles

We looked for extracellular vesicles with spherical shapes (20–200 nm) in the *L. fermentum* U-21 culture supernatant. Transmission electron microscopy (TEM) was used to detect the presence of spherical structures (extracellular vesicles) in the ultracentrifugation of the bacterial culture supernatant. We verified the presence of structures with an approximate diameter of 80 nm (red arrow) in the *L. fermentum* U-21 ultracentrifugation residues, as shown in Figure 4a. The results of TEM observations shown in Figure 4a exposed that *L. fermentum* U-21 generated vesicle-like structures (approximately 80 nm).

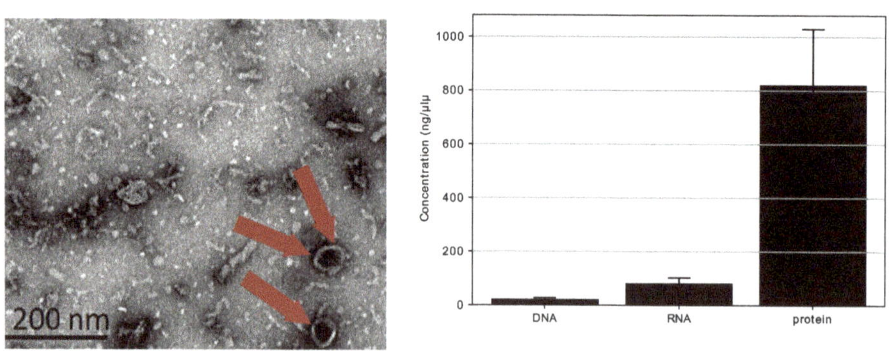

(**a**) (**b**)

Figure 4. Characterization of extracellular vesicles (red arrows) produced by *L. fermentum* U-21: (**a**) TEM images of *L. fermentum* U-21 extracellular vesicles. Red arrows indicate extracellular vesicles. Scale bar, 200 nm; (**b**) DNA, RNA, and protein in *L. fermentum* U-21 extracellular vesicles. Means ± SD, n = 3.

Additionally, protein (820.7 ng/µL) and RNA (81.9 ng/µL) were present in *L. fermentum* U-21 extracellular vesicles, whereas DNA (21.6 ng/µL) was rarely detected (Figure 4b). This result is similar to what was seen for *L. paracasei* [21]. To summarize, some lactobacilli release extracellular vesicles into culture supernatants. The extracellular vesicles of *L. fermentum* U-21 appear to have a cytoplasmic membrane and mainly contain bacterial proteins.

3.3. Isolation and Characteristic of the Metabolite Profile of the Extracellular Vesicles

In the extracellular vesicles of *L. fermentum* U-21, 360 metabolites were discovered, 82 of which are unique in comparison with *L. fermentum* 279 (Figure 5 and Supplementary Table S5).

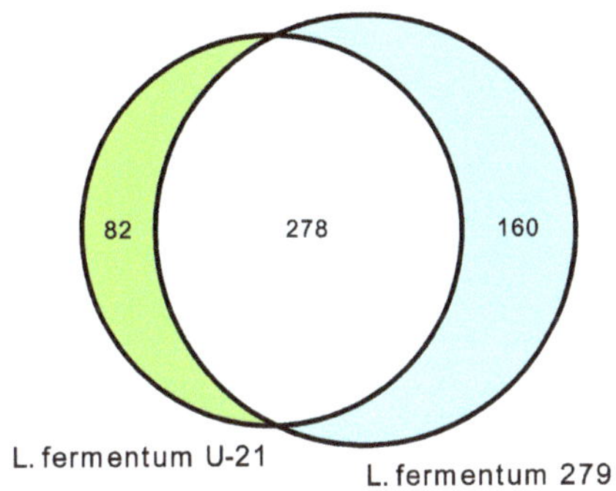

Figure 5. Venn diagrams show shared and unique metabolite features and their numbers for *L. fermentum* U-21 and *L. fermentum* 279 extracellular vesicles.

In the extracellular vesicles, forty-eight amino acids and their derivatives were found, six of which are unique to *L. fermentum* U-21: gamma-aminobutyric acid (GABA), cycloleucine, 3-hydroxy-tyrosine, alpha-methyl-L-tyrosine, aminomalonic acid, and alanylglycine. Extracellular vesicles included twenty-six sugars and carbohydrate conjugates, four of which are unique to *L. fermentum* U-21: 2,3,4-trihydroxybutyric acid, 2-keto-D-gluconic acid, α-D-mannopyranose, and arabinofuranose. There was also 2-keto-D-gluconic acid that is the key precursor for the synthesis of vitamin C detected. Nine metabolites were reported in the literature as bioactive compounds found in the extracellular vesicles, including GABA, adrenalin, dopamine, noradrenaline, melatonin, indole-3-acetamide, L-Norvaline, 4-Hydroxyphenyllactic acid with antioxidant properties, and penicillamine with anti-inflammatory properties (Table 5).

Table 5. Health beneficial compounds detected in the culture supernatant of *L. fermentum* U-21.

Metabolite	Class
Adrenaline	Neurotransmitter
GABA	
5-Hydroxytryptophan	Antidepressant
Timonacic	Antioxidant
5-Methoxytryptamine (5-MT)	
Indole-3-acetamide	Indoles and derivative
Melatonin	Hormone
2-keto-D-gluconic acid	Vitamin C precursor
L-Norvaline	Anti-inflammatory
Penicillamine	
3-Phelyllactic acid	Antimicrobial

Underlined metabolites are unique for *L. fermentum* U-21.

The sorting of metabolites by their metabolic pathways is performed with the metabolite set enrichment analysis (MSEA). Figure 6 shows results of the pathway enrichment analysis in MetaboAnalyst 5.0, which allow understanding of the main metabolic pathways.

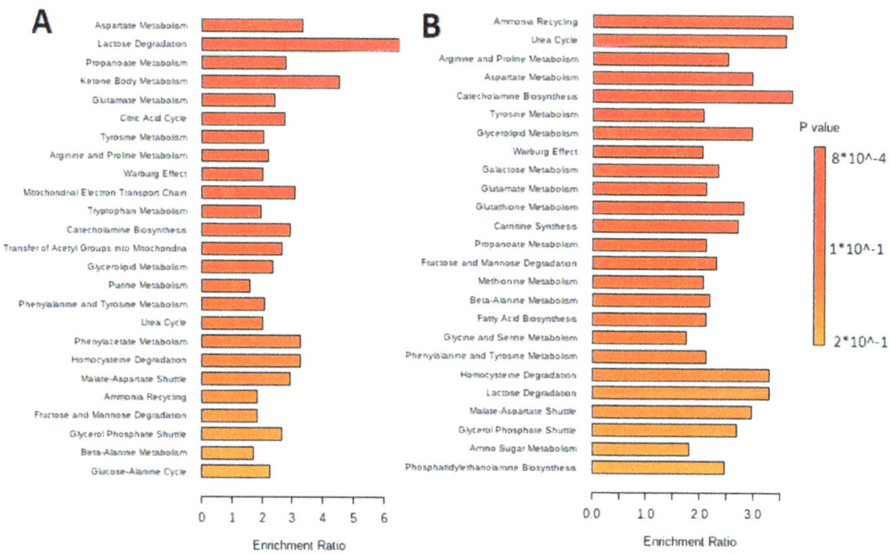

Figure 6. MetaboAnalyst 5.0 Pathway Enrichment Analysis for metabolites differed significantly between groups ($p < 0.05$). The enrichment ratio is calculated as the number of hits within a particular metabolic pathway divided by the expected number of hits. (**A**) Extracellular vesicles of *L. fermentum* U-21. (**B**) Extracellular vesicles of *L. fermentum* 279.

4. Discussion

Host health can be influenced by the human gut microbiota, which can have different microbial compositions and biochemical functions, so prebiotics, probiotics, and postbiotics have attracted the attention of scientists as a viable way to change the intestinal microbiome. A healthy GIT can be supported by probiotics, which are live microorganisms that, when administered in adequate amounts, confer a health benefit on the host [22] and typically consist of *Lactobacillus* and *Bifidobacterium* species. However, the postbiotic is a relatively new concept in medicine and science. Despite the fact that the idea of postbiotics is raised from the concept of producing the secretion of various metabolites with beneficial effects by the microbiota [23], the useful properties of postbiotics and their bioactivities, which determine these properties, are still unknown or unclear. In our laboratory, sequencing of the genome of the *L. fermentum* U-21 strain was previously carried out and 29 genes potentially determining its antioxidant activity were annotated in its genome [16]. In the same work, genes, whose expression increased in response to exposure to hydrogen peroxide, were identified [16].

This study aims to characterize an *L. fermentum* U-21 species as a potential postbiotic and identify distinct metabolites that determine the health-improving unique properties of *L. fermentum* U-21 by comparing the metabolic profiles of two Lactobacillus strains, one studied strain, *L. fermentum* U-21, and another strain, *L. fermentum* 279, which does not exhibit high AO properties.

While analyzing the untargeted metabolomic profile, a total of 1144 metabolites in the *L. fermentum* U-21′s cultural fluid were identified: 681 metabolites of them in the aqueous phase, 98 in the lipid-containing phase, and 365 in the extracellular vesicles, as amounts was expected from previous work [24]. While comparing the metabolites of the studied *L. fermentum* U-21 strain with the *L. fermentum* 279, which does not exhibit high

AO properties, 257 metabolites in the aqueous phase, 60 in the lipid-containing phase, and 82 in extracellular vesicles appeared to be unique for *L. fermentum* U-21. In addition, we identified that *L. fermentum* U-21 produced GABA, which is a nonprotein amino acid and a major inhibitory neurotransmitter in the mammalian central nervous system [25]. Many studies revealed considerable roles of GABA in different process such as modulating synaptic transmission, promoting neuronal development and relaxation, and preventing sleeplessness and depression [26–29]. Moreover, a recent study found that GABA has antihypertension, antidiabetes, anticancer, antioxidant, anti-inflammation, antimicrobial, and antiallergy properties [30]. However, the GABA synthesis gene has not been found in the genome of *L. fermentum* U-21, so we suppose that another metabolic pathway of GABA production is in this strain. Other neurotransmitters such as adrenaline, dopamine, noradrenaline, glycine, and L-DOPA were identified to be produced by *L. fermentum* U-21. An interesting fact is that a unique neurotransmitter found in *L. fermentum* U-21 but not *L. fermentum* 279 is L-DOPA. Since L-DOPA is the precursor of the neurotransmitters (dopamine, noradrenaline, and adrenaline), it is used in the therapeutic treatment of Parkinson's disease and dopamine-responsive dystonia in medicine [31]. The culture supernatant contained a number of short-chain fatty acids, including acetic acid, propanoic acid, and butanoic acid. These saturated aliphatic organics acids regulate the immune system response [32]. In particular, butyrate can inhibit histone deacetylases (HDAC) and the activation of nuclear factor kappa β (NF-kβ) in macrophages, which contribute to the immune and inflammatory response [33,34]. Some *Lactobacillus* species can convert tryptophan to indole, indole-3-aldehyde, and indole-3-lactic acid, which are signaling molecules to regulate epithelial integrity, immune response, and gastrointestinal motility through intestinal receptors. Indole and indole's derivative such as indolelactate, indole-3-acetamide, and indole-3-acetic acid produced by *L. fermentum* U-21 have been reported to display immunomodulatory benefits [35].

In our laboratory, the genes involved in the first of two glutathione, which is a critical endogenous antioxidant, synthesis steps and the transport of glutathione had been recently found in the genome of *L. fermentum* U-21 [9]. Whereas metabolomic analysis of the strain revealed pyroglutamic acid, ornithine, cysteine, glutamic acid, and glycine, from which glutathione is synthesized, and enhanced glutathione's synthesis [36].

The quantity of biological active metabolites of the promising postbiotic *Lactobacillus* strain *L. fermentum* U-21 was shown by this study of untargeted metabolomic analysis. This strain was firstly selected in an *E. coli* bioluminescent test-system for high AO potential [13]. *L. fermentum* 279, in turn, did not show high AO properties on this test-system. Previous studies have proved the high antioxidant and neuromodulation properties of *L. fermentum* U 21 by in vivo and in vitro models [13,14]. Moreover, recently unique genes coding the proteins involved in reducing antioxidant stress such as genes of a transcriptional regulator and genes responsible for heavy metal chelation and transportation were found in the *L. fermentum* U-21 and absence in the genome of *L. fermentum* 279 [16].

However, the biological compounds of the *L. fermentum* U-21 strain, which are responsible for such health-improving effects, have not been identified yet. Many strain-specific metabolites that can benefit the health of the host were found in this study. The production of these metabolites, which have positive health effects, may be the cause of this strain's postbiotic effect. Knowing particular metabolic pathways and the mechanisms of the generation of particular metabolites is helpful for understanding strain function in the GIT and host organism, as well as for understanding the standards for choosing particular postbiotics in the future. Complex omics studies, including metabolomics and proteomics analysis, and analysis of mRNA presence in vesicles of *L. fermentum* U-21 with regulatory function are the subject of our studies. Obviously, the impact of the complex metabolites, proteins, and mRNA contained in the vesicles and delivered to needful organs causes target effects of rare probiotic strains on specific diseases [37–39].

5. Conclusions

The unique metabolites produced by *L. fermentum* U-21, such as numerous organic acids, including fatty and amino acids, as well as a number of neurotransmitters and vitamins possess health-beneficial properties. Therefore, this strain can be used as an ingredient in pharmaceuticals and health food products, but further metabolomics studies are needed to precisely understand biologically active metabolites production's mechanisms. *L. fermentum* U-21 and its extracellular vesicles may be a potential candidate for the production of postbiotics and health food products.

Supplementary Materials: The following supporting information can be downloaded at: https://www.mdpi.com/article/10.3390/biotech12020039/s1. The files Supplementary 456BM.xlsx, Supplementary 789BM.xlsx and Supplementary 1011BM.xlsx contain the results, grouped samples were analyzed at a 1 to 50 flow split. The column "Sample"—contains the name of the sample, each sample was analyzed three times (4, 4_2, and 4_3 are technical repetitions of the sample). The samples 4, 5, 6 are the aqueous phases of *L. fermentum* U-21, *L. fermentum* 279, and MRS, respectively. The samples 7, 8, 9 are the lipid-containing phases of *L. fermentum* U-21, *L. fermentum* 279, and MRS, respectively. The samples 10 and 11 are the extracellular vesicles of *L. fermentum* U-21 and *L. fermentum* 279, respectively. The column "Name"—contains the name of the metabolite in the derivatized form, the RT column contains the retention times for both columns. The files Supplementary Supplementary.docx contain information about shared metabolites *of L. fermentum* U-21 and 279, and metabolites of *L. fermentum* 279 (Tables S1–S5).

Author Contributions: Conceptualization, A.A.V., Y.V.G. and V.N.D.; methodology, D.A.M., M.V.O., Y.V.G. and A.M.S.; formal analysis, Y.V.G.; investigation, D.A.M., M.V.O., Y.V.G. and A.M.S.; resources, A.A.V. and V.N.D.; writing—original draft preparation, A.A.V., Y.V.G. and V.N.D.; writing—review and editing, A.A.V., V.N.D. and R.A.I.; supervision, V.N.D.; project administration, A.A.V.; funding acquisition, V.N.D. All authors have read and agreed to the published version of the manuscript.

Funding: This research received no external funding.

Institutional Review Board Statement: Not applicable.

Informed Consent Statement: Not applicable.

Data Availability Statement: The analyzed open access materials are available online in issues on journal websites. The data are available in the Supplementary Materials.

Acknowledgments: We would like to thank the members of the Center for Collective Use "Human Proteome", at the facilities of which the data were obtained. TEM measurements were performed at the User Facilities Centre "Electron microscopy in life sciences" at Lomonosov Moscow State University.

Conflicts of Interest: The authors declare no conflict of interest.

References

1. Dettmer, K.; Aronov, P.A.; Hammock, B.D. Mass spectrometry-based metabolomics. *Mass Spectrom. Rev.* **2007**, *26*, 51–78. [CrossRef] [PubMed]
2. Cambeiro-Pérez, N.; Hidalgo-Cantabrana, C.; Moro-García, M.A.; Alonso-Arias, R.; Simal-Gándara, J.; Sánchez, B.; Martínez-Carballo, E. A metabolomics approach reveals immunomodulatory effects of proteinaceous molecules derived from gut bacteria over human peripheral blood mononuclear cells. *Front. Microbiol.* **2018**, *9*, 2701. [CrossRef] [PubMed]
3. Schell, M.A.; Karmirantzou, M.; Snel, B.; Vilanova, D.; Berger, B.; Pessi, G.; Zwahlen, M.C.; Desiere, F.; Bork, P.; Delley, M.; et al. The genome sequence of Bifidobacterium longum reflects its adaptation to the human gastrointestinal tract. *Proc. Natl. Acad. Sci. USA* **2002**, *99*, 14422–14427. [CrossRef]
4. Duar, R.M.; Lin, X.B.; Zheng, J.; Martino, M.E.; Grenier, T.; Pérez-Muñoz, M.E.; Leulier, F.; Gänzle, M.; Walter, J. Lifestyles in transition: Evolution and natural history of the genus *Lactobacillus*. *FEMS Microbiol. Rev.* **2017**, *41*, S27–S48. [CrossRef] [PubMed]
5. Dempsey, E.; Corr, S.C. *Lactobacillus* spp. for gastrointestinal health: Current and future perspectives. *Front. Immunol.* **2022**, *13*, 840245. [CrossRef]
6. Mishra, V.; Shah, C.; Mokashe, N.; Chavan, R.; Yadav, H.; Prajapati, J. Probiotics as potential antioxidants: A systematic review. *J. Agric. Food Chem.* **2015**, *63*, 3615–3626. [CrossRef]

7. Nowak, A.; Paliwoda, A.; Błasiak, J. Anti-proliferative, pro-apoptotic and anti-oxidative activity of *Lactobacillus* and Bifidobacterium strains: A review of mechanisms and therapeutic perspectives. *Crit. Rev. Food Sci. Nutr.* **2019**, *59*, 3456–3467. [CrossRef]
8. Danilenko, V.N.; Stavrovskaya, A.V.; Voronkov, D.N.; Gushchina, A.S.; Marsova, M.V.; Yamshchikova, N.G.; Ol'shansky, S.; Ivanov, M.; Illarioshkin, S.N. The use of a pharmabiotic based on the *Lactobacillus* fermentum U-21 strain to modulate the neurodegenerative process in an experimental model of Parkinson disease. *Ann. Clin. Exp. Neurol.* **2020**, *14*, 62–69.
9. Poluektova, E.; Yunes, R.; Danilenko, V. The putative antidepressant mechanisms of probiotic bacteria: Relevant genes and proteins. *Nutrients* **2021**, *13*, 1591. [CrossRef]
10. Zhou, Y.; Gong, W.; Xu, C.; Zhu, Z.; Peng, Y.; Xie, C. Probiotic assessment and antioxidant characterization of *Lactobacillus* plantarum GXL94 isolated from fermented chili. *Front. Microbiol.* **2022**, *13*, 997940. [CrossRef]
11. Thorakkattu, P.; Khanashyam, A.C.; Shah, K.; Babu, K.S.; Mundanat, A.S.; Deliephan, A.; Deokar, G.S.; Santivarangkna, C.; Nirmal, N.P. Postbiotics: Current trends in food and Pharmaceutical industry. *Foods* **2022**, *11*, 3094. [CrossRef] [PubMed]
12. Averina, O.V.; Poluektova, E.U.; Marsova, M.V.; Danilenko, V.N. Biomarkers and utility of the antioxidant potential of probiotic Lactobacilli and Bifidobacteria as representatives of the human gut microbiota. *Biomedicines* **2021**, *9*, 1340. [CrossRef] [PubMed]
13. Marsova, M.; Abilev, S.; Poluektova, E.; Danilenko, V. A bioluminescent test system reveals valuable antioxidant properties of *lactobacillus* strains from human microbiota. *World J. Microbiol. Biotechnol.* **2018**, *34*, 1–9. [CrossRef] [PubMed]
14. Marsova, M.; Poluektova, E.; Odorskaya, M.; Ambaryan, A.; Revishchin, A.; Pavlova, G.; Danilenko, V. Protective effects of *Lactobacillus* fermentum U-21 against paraquat-induced oxidative stress in Caenorhabditis elegans and mouse models. *World J. Microbiol. Biotechnol.* **2020**, *36*, 1–10. [CrossRef] [PubMed]
15. Noureen, S.; Riaz, A.; Arshad, M.; Arshad, N. In vitro selection and in vivo confirmation of the antioxidant ability of *Lactobacillus* brevis MG000874. *J. Appl. Microbiol.* **2019**, *126*, 1221–1232. [CrossRef]
16. Poluektova, E.; Mavletova, D.; Odorskaya, M.; Marsova, M.; Klimina, K.; Koshenko, T.; Yunes, R.; Danilenko, V. Comparative Genomic, Transcriptomic, and Proteomic Analysis of the Limosi*lactobacillus* fermentum U-21 Strain Promising for the Creation of a Pharmabiotic. *Russ. J. Genet.* **2022**, *58*, 1079–1090. [CrossRef]
17. Schramm, F.D.; Schroeder, K.; Jonas, K. Protein aggregation in bacteria. *FEMS Microbiol. Rev.* **2020**, *44*, 54–72. [CrossRef]
18. Alam, A.; Bröms, J.E.; Kumar, R.; Sjöstedt, A. The role of ClpB in bacterial stress responses and virulence. *Front. Mol. Biosci.* **2021**, *8*, 668910. [CrossRef]
19. Vermassen, A.; Leroy, S.; Talon, R.; Provot, C.; Popowska, M.; Desvaux, M. Cell wall hydrolases in bacteria: Insight on the diversity of cell wall amidases, glycosidases and peptidases toward peptidoglycan. *Front. Microbiol.* **2019**, *10*, 331. [CrossRef]
20. Fiehn, O. Metabolomics by gas chromatography–mass spectrometry: Combined targeted and untargeted profiling. *Curr. Protoc. Mol. Biol.* **2016**, *114*, 30–34. [CrossRef]
21. Choi, J.H.; Moon, C.M.; Shin, T.S.; Kim, E.K.; McDowell, A.; Jo, M.K.; Joo, Y.H.; Kim, S.E.; Jung, H.K.; Shim, K.N.; et al. Lactobacillus paracasei-derived extracellular vesicles attenuate the intestinal inflammatory response by augmenting the endoplasmic reticulum stress pathway. *Exp. Mol. Med.* **2020**, *52*, 423–437. [CrossRef] [PubMed]
22. Hill, C.; Guarner, F.; Reid, G.; Gibson, G.R.; Merenstein, D.J.; Pot, B.; Morelli, L.; Canani, R.B.; Flint, H.J.; Salminen, S.; et al. Expert consensus document: The International Scientific Association for Probiotics and Prebiotics consensus statement on the scope and appropriate use of the term probiotic. *Nat. Rev. Gastroenterol. Hepatol.* **2014**, *11*, 506–514. [CrossRef] [PubMed]
23. Chaudhari, A.; Dwivedi, M.K. The concept of probiotics, prebiotics, postbiotics, synbiotics, nutribiotics, and pharmabiotics. In *Probiotics in the Prevention and Management of Human Diseases*; Elsevier: Amsterdam, The Netherlands, 2022; pp. 1–11.
24. Chamberlain, M.; O'Flaherty, S.; Cobián, N.; Barrangou, R. Metabolomic Analysis of *Lactobacillus acidophilus, L. gasseri, L. crispatus*, and *Lacticaseibacillus rhamnosus* Strains in the Presence of Pomegranate Extract. *Front. Microbiol.* **2022**, *13*, 863228. [CrossRef] [PubMed]
25. Cui, Y.; Miao, K.; Niyaphorn, S.; Qu, X. Production of gamma-aminobutyric acid from lactic acid bacteria: A systematic review. *Int. J. Mol. Sci.* **2020**, *21*, 995. [CrossRef]
26. Stagg, C.J.; Bachtiar, V.; Johansen-Berg, H. The role of GABA in human motor learning. *Curr. Biol.* **2011**, *21*, 480–484. [CrossRef]
27. Nuss, P. Anxiety disorders and GABA neurotransmission: A disturbance of modulation. *Neuropsychiatr. Dis. Treat.* **2015**, *11*, 165–175.
28. Gottesmann, C. GABA mechanisms and sleep. *Neuroscience* **2002**, *111*, 231–239. [CrossRef]
29. Wagner, S.; Castel, M.; Gainer, H.; Yarom, Y. GABA in the mammalian suprachiasmatic nucleus and its role in diurnal rhythmicity. *Nature* **1997**, *387*, 598–603. [CrossRef]
30. Rashmi, D.; Zanan, R.; John, S.; Khandagale, K.; Nadaf, A. γ-aminobutyric acid (GABA): Biosynthesis, role, commercial production, and applications. *Stud. Nat. Prod. Chem.* **2018**, *57*, 413–452.
31. Ovallath, S.; Sulthana, B. Levodopa: History and therapeutic applications. *Ann. Indian Acad. Neurol.* **2017**, *20*, 185.
32. Nogal, A.; Valdes, A.M.; Menni, C. The role of short-chain fatty acids in the interplay between gut microbiota and diet in cardio-metabolic health. *Gut Microbes* **2021**, *13*, 1897212. [CrossRef] [PubMed]
33. Glauben, R.; Siegmund, B. Inhibition of histone deacetylases in inflammatory bowel diseases. *Mol. Med.* **2011**, *17*, 426–433. [CrossRef] [PubMed]
34. Lührs, H.; Gerke, T.; Müller, J.; Melcher, R.; Schauber, J.; Boxberger, F.; Scheppach, W.; Menzel, T. Butyrate inhibits NF-κB activation in lamina propria macrophages of patients with ulcerative colitis. *Scand. J. Gastroenterol.* **2002**, *37*, 458–466. [CrossRef] [PubMed]

35. Li, X.; Zhang, B.; Hu, Y.; Zhao, Y. New insights into gut-bacteria-derived indole and its derivatives in intestinal and liver diseases. *Front. Pharmacol.* **2021**, *12*, 3531. [CrossRef] [PubMed]
36. Gould, R.L.; Pazdro, R. Impact of supplementary amino acids, micronutrients, and overall diet on glutathione homeostasis. *Nutrients* **2019**, *11*, 1056. [CrossRef]
37. Kameli, N.; Borman, R.; López-Iglesias, C.; Savelkoul, P.; Stassen, F.R. Characterization of Feces-Derived Bacterial Membrane Vesicles and the Impact of Their Origin on the Inflammatory Response. *Front. Cell. Infect. Microbiol.* **2021**, *11*, 667987. [CrossRef]
38. Díez-Sainz, E.; Milagro, F.I.; Riezu-Boj, J.I.; Lorente-Cebrián, S. Effects of gut microbiota-derived extracellular vesicles on obesity and diabetes and their potential modulation through diet. *J. Physiol. Biochem.* **2022**, *78*, 485–499. [CrossRef]
39. Ahmadi Badi, S.; Bruno, S.P.; Moshiri, A.; Tarashi, S.; Siadat, S.D.; Masotti, A. Small RNAs in Outer Membrane Vesicles and Their Function in Host-Microbe Interactions. *Front. Microbiol.* **2020**, *11*, 1209. [CrossRef]

BioTech

Article

Assessing Curcumin Uptake and Clearance and Their Influence on Superoxide Dismutase Activity in *Drosophila melanogaster*

Tammy R. Hoffman [1,†], Sarah A. Emsley [1,†], Jenna C. Douglas [1], Kaela R. Reed [2], Abigail R. Esquivel [3], Marc J. Koyack [4], Brie E. Paddock [1] and Patrick Videau [1,*,‡]

1 Department of Biology, Southern Oregon University, Ashland, OR 97520, USA
2 Department of Chemistry, Southern Oregon University, Ashland, OR 97520, USA
3 Department of Molecular Medicine, Morsani College of Medicine, University of South Florida, Tampa, FL 33602, USA
4 School of Arts and Sciences, Gwynedd Mercy University, Gwynedd Valley, PA 19437, USA
* Correspondence: videaup@sou.edu; Tel.: +1-514-552-6344
† These authors contributed equally to the work.
‡ Current address: AgBiome, Research Triangle Park, NC 27709, USA.

Abstract: While normal levels of reactive oxygen and nitrogen species (RONS) are required for proper organismal function, increased levels result in oxidative stress. Oxidative stress may be managed via the scavenging activities of antioxidants (e.g., curcumin) and the action of enzymes, including superoxide dismutase (SOD). In this work, the uptake and clearance of dietary curcuminoids (consisting of curcumin, demethoxycurcumin, and bisdemethoxycurcumin) was assessed in *Drosophila melanogaster* larvae following chronic or acute exposure. High levels of curcuminoid uptake and loss were observed within a few hours and leveled off within eight hours post treatment onset. The addition or removal of curcuminoids from media resulted in corresponding changes in SOD activity, and the involvement of each of the three *SOD* genes was assessed for their contribution to total SOD activity. Taken together, these data provide insight into the uptake and clearance dynamics of curcuminoids and indicate that, while SOD activity generally increases following curcuminoid treatment, the individual *SOD* genes appear to contribute differently to this response.

Keywords: *Drosophila*; curcumin; superoxide dismutase

Key Contribution: This work provides insight into the uptake and clearance dynamics of curcumin and related compounds (curcuminoids), and assesses the influence of three superoxide dismutase isoforms in responding to curcuminoid treatment.

1. Introduction

Oxidative stress, the imbalance between reactive oxygen and nitrogen species (RONS) and endogenous antioxidant systems, has been implicated in disease [1]. RONS are a group of short lived and highly reactive species, including superoxide (O_2^-), peroxide (OH^-), nitrogen dioxide ($\bullet NO_2$), and others. Although normal RONS levels are necessary to maintain proper signaling and physiological functions, excessive amounts can lead to the denaturation and degradation of proteins and DNA, lipid peroxidation, toxicity, and cell death [2,3]. As research has indicated oxidative stress may be responsible for initiating and accelerating chronic and degenerative disease and aging, it is crucial to identify potential therapeutic compounds that directly target and reduce excessive RONS to physiological levels. In recent years, curcumin, a yellow phenolic compound derived from the *Curcuma longa* plant, has been studied for its multiple bioactivities, including its function as a potent antioxidant [4–6]. While it is speculated that the polyphenolic rings in curcumin may directly scavenge RONS, curcumin may also act as a signaling molecule that influences the expression and/or activity of enzymes involved in reducing oxidative stress and its effects. Although the mechanism by which curcumin acts remains under investigation, research indicates that curcumin treatment results in extensive changes to gene expression, specifically in genes related to endogenous antioxidants [5,7–10]. These results suggest that curcumin may be acting, at least in part, via modulation of specific protective genetic elements.

Previous studies have shown that curcumin treatment results in an increase in superoxide dismutase (SOD) activity, an enzyme that catalyzes the conversion of superoxide into oxygen (O_2) and hydrogen peroxide (H_2O_2) [2]. SOD exists in three different forms encoded by separate genes. While these forms have functional similarities, they are distinct in sequence, structure, cofactor requirements, and cellular compartmentalization [2,11]. SOD1 exists almost exclusively as a cytosolic 88 kDa homodimer that requires copper and zinc cofactors to function [12]. In contrast, SOD2 is a 32kDa homotetramer existing exclusively in the mitochondria, and has a manganese cofactor. Lastly, SOD3 is a 135kDa tetramer that primarily functions extracellularly, and utilizes copper and zinc cofactors. Changes in SOD activity and expression have also been linked to disease [2,13]. SOD has multiple genetic links to amyotrophic lateral sclerosis (ALS), a debilitating neurological disease [11,14,15], as well as links to chronic kidney disease, cancer progression, and chronic obstructive pulmonary disease (COPD) [16–18]. Disease intervention by targeting or mimicking elements within the antioxidant systems, such as SOD, has led to promising results in recent studies. In one study, the ingestion of cerium oxide nanoparticles ($nCeO_2$) correlated with the restoration of SOD levels in a transgenic *Drosophila melanogaster* Alzheimer's disease model [19]. As *Drosophila* is an accessible model system to study the molecular underpinnings of disease, strides have been taken to study SOD in this model to provide the basis for developing future therapies [20,21].

Previous studies have indicated a link between SOD expression and/or activity and responses to curcumin treatment. *Drosophila* fed a curcumin-supplemented diet were found to have an increased lifespan, which co-occurred with increased SOD activity and upregulation of SOD gene expression [9]. Similarly, feeding *Drosophila* larvae curcumin increased SOD and acetylcholinesterase activities while promoting an increased lifespan [22]. Under heat stress conditions, feeding curcumin to *Drosophila* increased lifespan and SOD gene expression [23]. The addition of disulfiram, an SOD inhibitor, negated the lifespan increases associated with curcumin-supplemented diets in *Drosophila* [24]. Curcumin treatment also mitigated both copper-induced oxidative stress and increased aging rates following irradiation in *Drosophila*, and oxidative stress in astrocytes [25–27]. Interestingly, in the aforementioned studies that identified increased SOD activity following curcumin treatment, only total SOD activity was assessed, and the publications did not distinguish between activities originating from individual SOD homologs.

We previously found that curcumin treatment in *Drosophila melanogaster* lowered the oxidative stress burden, increased lifespan, and affected behavior in a sex-dependent manner [28]. It was found that, when reared on several concentrations of curcuminoids, the larvae accumulated curcuminoids in their tissues and this was facilely quantified via methanol extraction and HPLC analysis. Additionally, several sex-specific outcomes were observed: increases in longevity and decreases in oxidative stress in females, as well as motility issues in males. While formative, this study focused on phenotypic sex differences and did not elucidate a specific mechanism underlying the decreased oxidative stress levels observed. Given the differences among SOD homologs, understanding how curcumin may influence each will begin to define a mechanism by which curcumin may act to modulate oxidative stress. Here, we assess the uptake and clearance of dietary curcumin in *Drosophila* and determine its influence on modulating isoform-specific SOD activity.

2. Materials and Methods

2.1. Drosophila Husbandry and Curcuminoid Treatments

All *Drosophila melanogaster* were maintained as previously described [28]. Briefly, all strains were maintained on Nutri-Fly German formulation medium (GF; Genessee Scientific,

San Diego, CA, USA) at 25 °C in a 12 h:12 h light:dark cycle at 60% humidity. The Oregon R strain is denoted as wild type (WT) in this work. The SOD1, SOD2, and SOD3 deficiency lines utilized, Sod1[X-39] e[1]/TM3, Sb[1] Ser[1] (#24490), Df(2R)Exel7145 (#7887), and y[1] w[67c23] (#14138), respectively, were obtained from the Bloomington Stock Center and are referred to as SOD1, SOD2, and SOD3 throughout. These were chosen because the individual SOD mutations each line harbors have been the focus of previous studies and represent useful lines for assessment [29–31]. Curcuminoid treatments were prepared with the desired concentration of curcuminoids (0–25 mM final concentrations in GF medium), derived from *Curcuma longa* (C1386, Sigma-Aldrich, St Louis, MO, USA; 98% pure standard consisting of curcumin, demethoxycurcumin, and bisdemethoxycurcumin in a 67:29:4 ratio and referred to as curcuminoids hereafter) as previously described [28].

2.2. *Larval Preparation, Extraction, and HPLC Analysis*

Groups of 10 third-instar larvae were collected and extracted as previously described [28]. Larvae reared on GF medium or GF supplemented with 25 mM curcuminoids were rinsed in 1× PBS and blotted dry before transfer onto a new medium for the durations stated. HPLC analysis of curcuminoid levels in larvae was conducted as previously described [28]. To assess curcuminoid levels that could enter an SOD assay (described below), larvae were extracted with extraction buffer (10 mM Tris-base pH 7.4) and the rest of the protocol and analysis remained unchanged.

2.3. *Larval Preparation and SOD Assay*

Larvae were reared from hatching on GF medium or GF supplemented with 25 mM curcuminoids, and were rinsed in 1× PBS and blotted dry before use. All manipulations were conducted on ice within 30 min of harvest. The 1 h PBS incubation conducted prior to extraction for HPLC analysis was omitted to ensure that the SOD activity levels measured were not altered. After rinsing, 250 µL of extraction buffer (10 mM Tris-base pH 7.4) was added to groups of 5 larvae, which were ground for roughly 1 min using a motorized Kontes pellet pestle rod system and then sonicated on ice using a Qsonica Q125 sonicator (Melville, NY, USA) with a 0.25-inch probe for one pulse of 10 s at 40% power. Extracts were centrifuged at 14,000× *g* for 10 min and the supernatant was transferred to a fresh tube, taking care not to disturb the pellet. Protein concentrations were determined via Bradford assay using the Bio-Rad Protein Assay Kit II according to the manufacturer's instructions (Bio-Rad, Hercules, CA, USA). SOD activity was quantified using the Superoxide Dismutase (SOD) Activity Assay Kit according to the manufacturer's instructions (#CS0009; Sigma-Aldrich, St. Louis, MO, USA). Activity was measured in technical duplicate for each biological replicate, and the results were averaged from the means of ten biological replicates per treatment.

2.4. *Analyses*

Significance in curcuminoid concentrations and SOD activities between conditions was assessed via student's *t*-test and an alpha value of 0.05, which were conducted with the Microsoft Excel software (Microsoft Corporation, Redmond, WA, USA). Graphical preparation of data was conducted using the GraphPad Prism software (GraphPad, San Diego, CA, USA).

3. Results

3.1. *Assessing Uptake and Clearance Parameters of Curcuminoids in Drosophila Larvae*

While our previous study demonstrated quantitative methodology and chemical, developmental, and behavioral changes that were altered in crawling third-instar WT *Drosophila melanogaster* larvae following curcuminoid (curcumin and its derivatives demethoxycurucmin and bisdemethoxycurcumin) treatment, additional relevant parameters remain undefined in this model system [28]. Larvae reared on either 0.25, 2.5, or 25 mM curcuminoids retained different levels of curcuminoids, but the influence of curcuminoid concentration on tissue uptake and retention is unknown. To address this, *Drosophila* larvae were reared from hatching on various curcuminoid concentrations (a chronic treatment) and the level of curcuminoids in their tissues was quantified in seven-day wandering third-instar larvae. The level of curcuminoids in larval tissue increased concomitant with curcuminoid concentration in the medium and began to plateau by 15 mM curcuminoids in media (Figure 1A). The average curcuminoid concentrations in tissues of larvae reared on 15, 20, and 25 mM curcuminoids were 274 ± 44, 262 ± 45, and 315 ± 52 ng/mg of tissue, respectively, which were not significantly different from one another (paired *t*-tests, each $p > 0.41$).

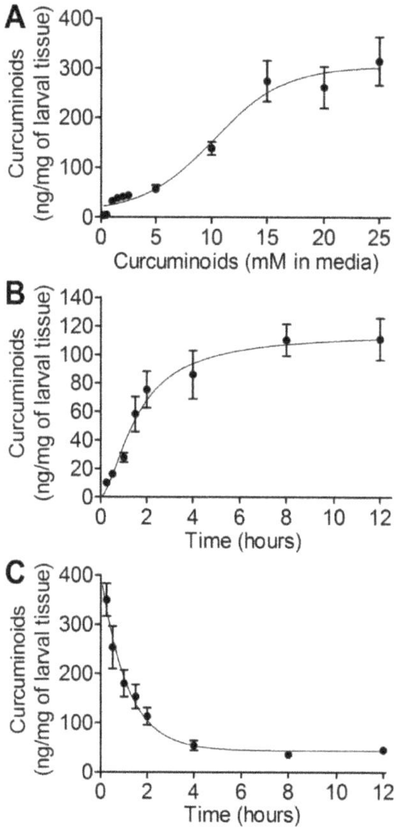

Figure 1. Determination of the influence of time and concentration on curcuminoid uptake in wild type (WT) larvae. Larvae were reared for seven days and the level of curcuminoids in tissues was quantified from groups of 10 crawling third-instar larvae ($n = 10$ per treatment). Larvae were reared on varying curcuminoid concentrations (**A**) on 0 mM curcuminoids and transferred to 25 mM curcuminoids (**B**), or on 25 mM curcuminoids and transferred to 0 mM curcuminoids (**C**), which resulted in the following equations of best fit: $y = 10.23 + \frac{294.17}{1+10^{10.26-x}}$, ($r^2 = 0.7382$); $y = \frac{115.4*X^{1.608}}{2.151+X^{1.608}}$, ($r^2 = 0.7650$); and $y = 359.38^{-0.885*X} + 43.42$, ($r^2 = 0.7562$), respectively. The mean of experiments conducted in $n = 10$ is presented, and error bars represent the standard error of the mean.

The results presented above provide insight into the bounds of curcuminoid carrying capacity as a result of chronic exposure to curcuminoids but do not describe the time-dependence of curcuminoid uptake or loss (an acute exposure). To address this, larvae reared on 0 or 25 mM curcuminoids were transferred to the opposite condition (i.e., larvae reared on 0 mM were transferred to 25 mM curcuminoids and vice versa) and curcuminoids in their tissues were quantified over the next 12 h. Curcuminoid uptake was quick in the first two hours immediately following larval transfer onto media supplemented with 25 mM curcuminoids, and began to stabilize roughly four hours post transfer (Figure 1B). The highest concentration of curcuminoids reached after 12 h of exposure was 111 ± 16 ng/mg of tissue, which is significantly lower than the concentration reached after seven days of chronic exposure (315 ± 52 ng/mg of tissue; *t*-test, $p = 0.003$). Similarly, the transition from 25 mM to 0 mM curcuminoids saw a swift drop in curcuminoid concentration in the tissues, which began to level off about four hours post transfer (Figure 1C). After 12 h on medium lacking curcuminoids, the larvae maintained 45 ± 7 ng/mg of tissue, which is similar to the level seen after seven days on medium supplemented with 5 mM curcuminoids (57 ± 9 ng/mg of tissue; *t*-test, $p = 0.27$). Taken together, these results indicate that curcuminoids display uptake and clearance within hours of treatment onset and that curcuminoid levels are retained and cleared differently between acute and chronic exposures.

3.2. SOD Levels Are Modulated in Response to Curcuminoid Treatment

Previous research has indicated that dietary curcumin decreases oxidative stress levels and leads to increased longevity [32,33]. Decreases in the levels of oxidative stress markers coincide with an increase in SOD expression, which would directly influence metrics of oxidative stress. While the link between dietary curcumin exposure and increased SOD expression has been demonstrated, the temporal dynamics of this interaction remain unexplored. To address this, a commercial SOD assay was used to measure SOD activity following chronic curcuminoid exposure and the addition or removal of curcuminoids from larval diets (Figure 2A). Following protein extraction from tubes of five washed and homogenized larvae (constituting a single biological replicate), it was found that 1 µg of total protein was sufficient to provide reproducible data. SOD activity in total protein from larvae reared without curcuminoids was measured immediately after extraction (0.79 ± 0.26 units/µg of protein) and following incubation on ice for one hour (0.66 ± 0.27 units/µg of protein). This duration on ice was within the time frame in which SOD activity was measured after extraction, and the additional time did not significantly alter SOD activity in samples (*t*-test, $p = 0.7214$). The SOD assay relies on the generation of superoxides by xanthine/xanthine oxidase for conversion into hydrogen peroxide and water by SOD; this reaction could be susceptible to the additional radical scavenging activity of curcuminoids present in larval tissue. Larvae reared in 25 mM curcuminoids were extracted with protein extraction buffer and HPLC analysis showed that extracted curcuminoids were below the limit of detection (1 ng on column or 45 nM). The addition of 0.01–50 nM curcuminoids to SOD assays using pure SOD supplied with the SOD assay kit did not result in significant decreases in activity at any concentration tested (Figure 2B; pairwise *t*-test, $p > 0.07$). The optimization steps described above provided the basis for the reproducible quantification of SOD activity in larval total protein extracts.

Studies have indicated that *Drosophila* reared on curcumin exhibit increased levels of SOD activity [9,22]. Quantification of the SOD activity of larvae reared on 25 mM curcuminoids (2.6 ± 0.9 units/µg of protein) displayed a 2.7-fold higher average SOD activity than larvae reared without curcuminoid supplementation (Figure 3A; 1.0 ± 0.2 units/µg of protein). In these experiments, SOD activity is measured after chronic exposure to curcumin; however, the influence of acute curcuminoid exposure or the removal of curcuminoids on SOD activity remained to be investigated. To determine the activity of SOD in response to acute curcuminoid exposure or the removal of curcuminoids, larvae reared on 0 or 25 mM curcuminoids were transferred to the opposite condition as described above, and SOD activity was quantified over the following 12 h. Larvae reared on 0 mM curcuminoids (1.0 ± 0.2 units/µg of protein) and transferred to 25 mM curcuminoids had a roughly 6.9-fold increase in average SOD activity after 2 h of exposure (6.7 ± 1.4 units/µg of protein). This level of SOD activity was consistent up to 12 h of exposure (Figure 3B). Interestingly, the SOD activity level measured following acute curcuminoid exposure is

2.4-fold higher on average than the level observed following seven days of chronic exposure to curcuminoids. In a similar trend, larvae reared on 25 mM curcuminoids and transferred to 0 mM curcuminoids saw a roughly 3.5-fold decrease in average SOD activity 2 h after transfer, which returned to the baseline level found in larvae raised on 0 mM curcuminoids. This level of SOD activity remained up to 12 h of exposure. While it is possible that the act of transferring larvae would be stressful and result in increases in SOD activity, transferring larvae from 0 mM curcuminoids to new medium with 0 mM curcuminoids did not alter SOD activity measurably. Taken together, the results indicate that the modulation of SOD activity occurs concomitantly with fluctuating curcuminoid levels. We infer that the changes in tissue curcuminoid levels directly or indirectly influence the mirrored trends in SOD activities.

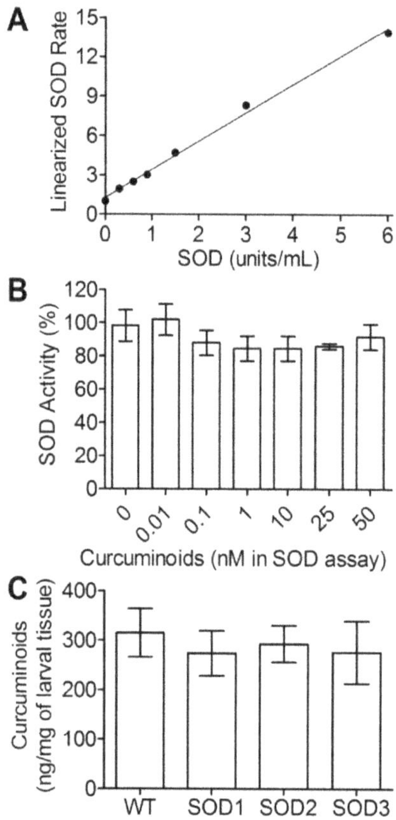

Figure 2. Parameters of the SOD activity assay used herein. A standard curve was prepared with purified SOD (**A**) with an equation of best fit: $y = 2.156 * X + 1.274$, ($r^2 = 0.9945$). The addition of varying concentrations of curcuminoids to the SOD assay, conducted with a constant level of 3 units/mL SOD, is expressed as a percentage of the activity values recorded without the addition of curcuminoids (**B**). Each of three replicates was conducted in technical duplicate, averaged, and the mean of experiments conducted in triplicate is presented (**A**,**B**). Larvae from all strains were reared for seven days on 25 mM curcuminoids and the level of curcuminoids in tissues was quantified from groups of 10 crawling third-instar larvae (**C**) (n = 10 per treatment). The mean of experiments conducted in n = 10 is presented. Error bars in all panels represent the standard error of the mean. No significant differences were found in the data presented in panels (**B**,**C**).

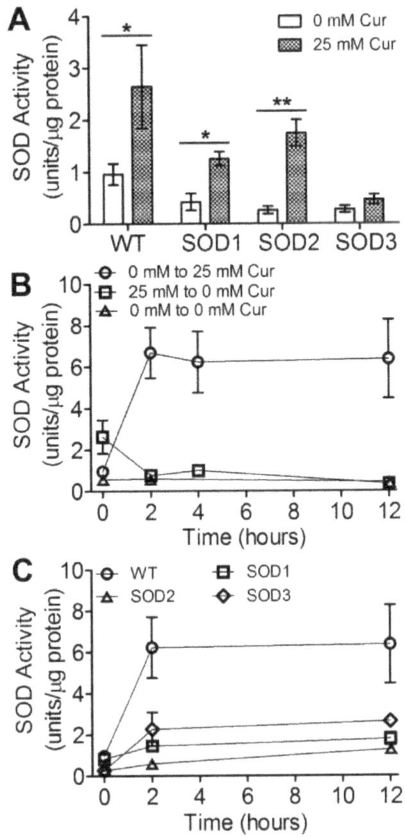

Figure 3. The influence of chronic and acute exposure to curcuminoids on SOD activity in the wild type (WT) and three SOD mutants. Larvae from all strains were reared for seven days on 0 or 25 mM curcuminoids; total protein was extracted, and SOD activity was quantified from groups of 5 crawling third-instar larvae (**A**) ($n = 5$ per treatment). WT larvae were reared for seven days on 0 or 25 mM curcuminoids, transferred to 25 mM or 0 mM (the opposite condition) for 2–12 h, total protein was extracted, and SOD activity was quantified from groups of 5 crawling third-instar larvae (**B**) ($n = 5$ per treatment). Larvae from all strains were reared for seven days on 0 curcuminoids, transferred to 25 mM for 2–12 h, total protein extracted, and SOD activity was quantified from groups of 5 crawling third-instar larvae (**C**) ($n = 5$ per treatment). The mean of experiments conducted in $n = 5$ is presented, and error bars represent the standard error of the mean; * $p < 0.05$, ** $p < 0.005$.

3.3. Three SOD Genes Differentially Contribute to Curcuminoid-Induced Activity

Previous work has identified three distinct genes encoding SOD homologs in *Drosophila*, each with differing localization profiles in the cell. Research on mechanisms underlying cellular responses to curcumin in *Drosophila* has investigated SOD activity in total protein but has not assessed the contribution of the individual *SOD* genes. To determine the individual influence of each of the three *Drosophila SOD* genes in curcuminoid-induced SOD activity, the SOD activity in lines mutant for each of the three *SOD* genes was assessed following chronic or acute curcuminoid exposure as conducted above. Consistent with published results, each SOD mutant strain reared on GF medium without curcuminoid supplementation displayed decreased SOD activity compared to the WT (Figure 3A). Rearing the mutant strains on 25 mM curcuminoids for seven days resulted in increased SOD activity from all three strains compared to the 0 mM curcuminoid treatment. Only the SOD2 mutant approached WT activity levels on 25 mM curcuminoids while the other mutants displayed lowered SOD activities. While it is possible that the multiple mutations in the SOD mutant lines tested could impact curcuminoid accumulation in tissues, curcuminoid levels in larvae reared on 25 mM curcuminoids were not significantly different from the WT (Figure 2C; pairwise *t*-tests, $p > 0.1$). These results indicate that 25 mM curcuminoid exposure increases SOD activity, and that the three *SOD* genes exhibit different levels of involvement in this response, with SOD3 being the most active in this treatment.

In the wild type data presented above, it became clear that SOD activity was greater during acute curcuminoid exposures than following chronic exposure. It is therefore possible that the results of chronic curcuminoid exposure result in diminished SOD levels in the three *SOD* mutants compared to acute treatments. To determine whether an increased SOD activity level was observed in response to acute curcuminoid exposure, larvae from each of the three *SOD* mutants were transferred to media with 25 mM curcuminoids, and SOD activity was measured up to 12 h post transfer (Figure 3C). Like the wild type, an increase in SOD activity was evident 2 h post transfer; however, while the *SOD1* and *SOD3* mutants attain their maximal SOD level 2 h post transfer, the *SOD2* mutant sees an increase in SOD activity up to 12 h post transfer. The SOD activity of the *SOD1* and *SOD3* mutants 12 h post transfer was higher than that observed following chronic curcuminoid exposure (*t*-test, $p < 0.05$). In contrast, the SOD activity of the *SOD2* mutant was not significantly different between the acute and chronic treatments (*t*-test, $p > 0.05$). These results indicate that each SOD gene is involved in the response to curcuminoid treatment, but the dynamics of this response differ among genes and between acute and chronic exposures.

4. Discussion

In this study, we tested the hypothesis that the antioxidant role of dietary curcuminoids involves increases in SOD activity. We demonstrated saturation in curcuminoid treatments beginning at about 15 mM in our model, and that there is rapid uptake and clearance of curcuminoids in *Drosophila* larvae exposed to or removed from curcuminoid-supplemented medium. We additionally found that both acute and chronic curcuminoid treatment results in increased total SOD activity, and differences in SOD activity were observed between acute and chronic treatments. We found that mutants of each of the three *SOD* gene homologs displayed lower total SOD activity at baseline, and that this level increases following treatment with curcuminoids. This study indicates that each *SOD* gene is involved in the antioxidant response to curcuminoid treatment and highlights the differences in homolog involvement.

While recent studies have demonstrated that curcumin treatments are associated with increased SOD levels in some models, most report on only one of the isoforms or on the total SOD response [7,8]. To determine the contribution of various SOD isoforms to the curcuminoid-induced SOD activity, three different SOD mutant lines were exposed to curcuminoids in both chronic and acute regiments. The three isoforms differ in cellular localization and transcriptional regulators [2]. Interestingly, all three deficiency lines showed less induction of SOD activity by curcuminoids than the wild type in the acute regimen, but none were significantly different from the other SOD deficiency lines. The SOD2 deficiency line demonstrated more SOD activity induction by curcuminoids in the chronic regimen compared to the other deficiency lines (Figure 3A), which may result from the heterozygosity of the SOD2 [31,34].

The specific mechanism of action by which curcuminoids influence SOD activity remains unknown, and research has shown that many genes are differentially regulated in response to curcumin treatment [35]. Interestingly, studies in cell lines and other model systems found that curcumin, delivered in media or in diets, altered the expression of many genes in multiple pathways (e.g., SOD, catalase, interleukins, antimicrobial peptides, growth factors, etc.), particularly via the activities of NF-κB and the nuclear factor erythroid 2-related factor 2 (Nrf2) regulator [7,36,37]. Nrf2 has been shown to regulate the expression of many genes involved in the response to oxidative stress, including SOD [38,39]. Our findings indicate that SOD activity responds to curcuminoid levels within two hours of acute treatments, which is a relatively quick response if it is mediated through changes in the expression of Nrf2 and/or other genes. Previous work on responses to various physiological changes and stress indicate that sweeping changes in gene expression have been recorded 10–60 min following treatment onset [40–42]. Given that SOD activity is an important component of the flies' response to acute oxidative stress, having a relatively quick response to treatment would seem beneficial for this organism.

Two different regimes of curcuminoid uptake were described in this work: a chronic condition where the larvae had curcuminoids in their diet throughout development and an acute condition where curcuminoids were only available for up to 12 h of feeding. Larvae, following chronic exposure, retain roughly three times more curcuminoids than larvae in the acute condition (Figure 1). Additionally, larvae reared in the chronic exposure treatment regimen initiated curcuminoid exposure as first-instar larvae, while those with acute exposure initiated treatment as third-instar larvae, which may underlie the observed differences in magnitude of SOD activity between the two exposure regimens. While the 25 mM concentration of curcuminoids is directly comparable between conditions, the developmental phase(s) were quite different. It is possible that curcuminoid uptake in first- and second-instar larvae is different than in third-instar larvae; the number of days of feeding somehow facilitates greater storage of curcuminoids. It is presumed that the curcuminoids are degraded at some rate within the larvae, so it is possible that degradation rates could be lower in earlier instar stages to allow greater levels of curcuminoids to accumulate. When larvae from the chronic curcuminoid condition were transferred to medium-lacking curcuminoids to assess clearance, some level of curcuminoids was retained within the larvae throughout the duration of the experiment. As the gut contents were allowed to clear for an hour prior to extraction, it is unlikely that this represented a large amount of the curcuminoids retained by larvae. It is possible that the first few hours of swift curcuminoid loss were due to clearance by the Malpighian tubules and the remaining curcuminoids were retained in fat globules and other tissues. While the above suggestions on possible underlying causes for the observed uptake and release of curcuminoids from larvae are strictly speculation, they will be the subject of future experiments.

5. Conclusions

In this work, the dynamics of curcuminoid uptake and clearance were assessed in *Drosophila* larvae and alterations in the level of these compounds appeared to mirror changes in SOD activity. Each of the three *SOD* gene homologs contributed to the curcuminoid-induced observed changes in SOD activity. As SOD activity is generally presented as total activity measurements without consideration of the influence of each *SOD* gene, this work provides insights into differential contributions to activity in response to curcuminoids. Presenting the initial measurements of dietary curcuminoid uptake and clearance and assessing the corresponding changes in SOD activity provide a general framework for consideration of the interplay of antioxidants and oxidative stress responses in this system.

Author Contributions: Conceptualization, P.V. and B.E.P.; methodology, T.R.H., S.A.E., M.J.K., B.E.P. and P.V.; formal analysis, P.V., M.J.K. and B.E.P.; investigation, T.R.H., S.A.E., M.J.K., J.C.D., K.R.R., A.R.E. and P.V.; writing—original draft preparation, P.V., B.E.P. and A.R.E.; writing—review and editing, all authors; funding acquisition, B.E.P., P.V. and M.J.K. All authors have read and agreed to the published version of the manuscript.

Funding: This work was supported by startup funds to B.E.P. and P.V. from Southern Oregon University and M.J.K. from Gwynedd Mercy University, and by a Goldwater Scholarship to S.A.E.

Institutional Review Board Statement: Not applicable.

Informed Consent Statement: Not applicable.

Data Availability Statement: The data used in this work are presented in their entirety.

Conflicts of Interest: The authors declare no conflict of interest. The funders had no role in the design of the study; in the collection, analyses, or interpretation of data; in the writing of the manuscript; or in the decision to publish the results.

References

1. Forman, H.J.; Zhang, H. Targeting oxidative stress in disease: Promise and limitations of antioxidant therapy. *Nat. Rev. Drug Discov.* **2021**, *20*, 689–709. [CrossRef] [PubMed]
2. Miao, L.; St Clair, D.K. Regulation of superoxide dismutase genes: Implications in disease. *Free Radic. Biol. Med.* **2009**, *47*, 344–356. [CrossRef] [PubMed]
3. Nguyen, T.T.; Vuu, M.D.; Huynh, M.A.; Yamaguchi, M.; Tran, L.T.; Dang, T.P.T. Curcumin effectively rescued Parkinson's disease-like phenotypes in a novel *Drosophila melanogaster* model with dUCH knockdown. *Oxid. Med. Cell. Longev.* **2018**, *2018*, 2038267. [CrossRef] [PubMed]
4. Jaroonwitchawan, T.; Chaicharoenaudomrung, N.; Namkaew, J.; Noisa, P. Curcumin attenuates paraquat-induced cell death in human neuroblastoma cells through modulating oxidative stress and autophagy. *Neurosci. Lett.* **2017**, *636*, 40–47. [CrossRef]
5. Han, J.; Pan, X.-Y.; Xu, Y.; Xiao, Y.; An, Y.; Tie, L.; Pan, Y.; Li, X.-J. Curcumin induces autophagy to protect vascular endothelial cell survival from oxidative stress damage. *Autophagy* **2012**, *8*, 812–825. [CrossRef]
6. Xu, X.-Y.; Meng, X.; Li, S.; Gan, R.-Y.; Li, Y.; Li, H.-B. Bioactivity, health benefits, and related molecular mechanisms of curcumin: Current progress, challenges, and perspectives. *Nutrients* **2018**, *10*, 1553. [CrossRef]
7. Lin, X.; Bai, D.; Wei, Z.; Zhang, Y.; Huang, Y.; Deng, H.; Huang, X. Curcumin attenuates oxidative stress in RAW264.7 cells by increasing the activity of antioxidant enzymes and activating the Nrf2-Keap1 pathway. *PLoS ONE* **2019**, *14*, e0216711. [CrossRef]
8. Di Tu, Q.; Jin, J.; Hu, X.; Ren, Y.; Zhao, L.; He, Q. Curcumin improves the renal autophagy in rat experimental membranous nephropathy via regulating the PI3K/AKT/mTOR and Nrf2/HO-1 signaling pathways. *Biomed. Res. Int.* **2020**, *2020*, 7069052. [CrossRef]
9. Shen, L.-R.; Xiao, F.; Yuan, P.; Chen, Y.; Gao, Q.-K.; Parnell, L.D.; Meydani, M.; Ordovas, J.M.; Li, D.; Lai, C.-Q. Curcumin-supplemented diets increase superoxide dismutase activity and mean lifespan in *Drosophila*. *AGE* **2013**, *35*, 1133–1142. [CrossRef]
10. Huang, L.; Chen, C.; Zhang, X.; Li, X.; Chen, Z.; Yang, C.; Liang, X.; Zhu, G.; Xu, Z. Neuroprotective effect of curcumin against cerebral ischemia-reperfusion via mediating autophagy and inflammation. *J. Mol. Neurosci.* **2018**, *64*, 129–139. [CrossRef]
11. Berdyński, M.; Miszta, P.; Safranow, K.; Andersen, P.M.; Morita, M.; Filipek, S.; Żekanowski, C.; Kuźma-Kozakiewicz, M. SOD1 mutations associated with amyotrophic lateral sclerosis analysis of variant severity. *Sci. Rep.* **2022**, *12*, 103. [CrossRef] [PubMed]
12. Zelko, I.N.; Mariani, T.J.; Folz, R.J. Superoxide dismutase multigene family: A comparison of the CuZn-SOD (SOD1), Mn-SOD (SOD2), and EC-SOD (SOD3) gene structures, evolution, and expression. *Free Radic. Biol. Med.* **2002**, *33*, 337–349. [CrossRef] [PubMed]
13. Fukai, T.; Ushio-Fukai, M. Superoxide dismutases: Role in redox signaling, vascular function, and diseases. *Antioxid. Redox Signal.* **2011**, *15*, 1583–1606. [CrossRef] [PubMed]
14. Zarei, S.; Carr, K.; Reiley, L.; Diaz, K.; Guerra, O.; Altamirano, P.F.; Pagani, W.; Lodin, D.; Orozco, G.; Chinea, A. A comprehensive review of amyotrophic lateral sclerosis. *Surg. Neurol. Int.* **2015**, *6*, 171. [CrossRef]
15. Valentine, J.S.; Hart, P.J. Misfolded CuZnSOD and amyotrophic lateral sclerosis. *Proc. Natl. Acad. Sci. USA* **2003**, *100*, 3617–3622. [CrossRef]
16. Dhar, S.K.; Tangpong, J.; Chaiswing, L.; Oberley, T.D.; St Clair, D.K. Manganese superoxide dismutase is a p53-regulated gene that switches cancers between early and advanced stages. *Cancer Res.* **2011**, *71*, 6684–6695. [CrossRef]
17. Du, Y.; Zhang, H.; Xu, Y.; Ding, Y.; Chen, X.; Mei, Z.; Ding, H.; Jie, Z. Association among genetic polymorphisms of GSTP1, HO-1, and SOD-3 and chronic obstructive pulmonary disease susceptibility. *Int. J. Chron. Obs. Pulmon. Dis.* **2019**, *14*, 2081–2088. [CrossRef]
18. Kitada, M.; Xu, J.; Ogura, Y.; Monno, I.; Koya, D. Manganese superoxide dismutase dysfunction and the pathogenesis of kidney disease. *Front. Physiol.* **2020**, *11*, 755. [CrossRef]
19. Sundararajan, V.; Venkatasubbu, G.D.; Sheik Mohideen, S. Investigation of therapeutic potential of cerium oxide nanoparticles in Alzheimer's disease using transgenic *Drosophila*. *3 Biotech.* **2021**, *11*, 159. [CrossRef]
20. Layalle, S.; They, L.; Ourghani, S.; Raoul, C.; Soustelle, L. Amyotrophic Lateral Sclerosis Genes in *Drosophila melanogaster*. *Int. J. Mol. Sci.* **2021**, *22*, 904. [CrossRef]
21. Surguchov, A. Invertebrate models untangle the mechanism of neurodegeneration in Parkinson's disease. *Cells* **2021**, *10*, 407. [CrossRef] [PubMed]
22. Akinyemi, A.J.; Oboh, G.; Ogunsuyi, O.; Abolaji, A.O.; Udofia, A. Curcumin-supplemented diets improve antioxidant enzymes and alter acetylcholinesterase genes expression level in *Drosophila melanogaster* model. *Metab. Brain Dis.* **2018**, *33*, 369–375. [CrossRef] [PubMed]
23. Chen, Y.; Liu, X.; Jiang, C.; Liu, L.; Ordovas, J.M.; Lai, C.-Q.; Shen, L. Curcumin supplementation increases survival and lifespan in *Drosophila* under heat stress conditions. *BioFactors* **2018**, *44*, 577–587. [CrossRef] [PubMed]
24. Suckow, B.K.; Suckow, M.A. Lifespan extension by the antioxidant curcumin in *Drosophila melanogaster*. *Int. J. Biomed. Sci.* **2006**, *2*, 402–405.

25. Seong, K.M.; Yu, M.; Lee, K.S.; Park, S.; Jin, Y.W.; Min, K.J. Curcumin mitigates accelerated aging after irradiation in *Drosophila* by reducing oxidative stress. *Biomed. Res. Int.* **2015**, *2015*, 425380. [CrossRef]
26. Daverey, A.; Agrawal, S.K. Curcumin alleviates oxidative stress and mitochondrial dysfunction in astrocytes. *Neuroscience* **2016**, *333*, 92–103. [CrossRef]
27. Abolaji, A.O.; Fasae, K.D.; Iwezor, C.E.; Aschner, M.; Farombi, E.O. Curcumin attenuates copper-induced oxidative stress and neurotoxicity in *Drosophila melanogaster*. *Toxicol. Rep.* **2020**, *7*, 261–268. [CrossRef]
28. Esquivel, A.R.; Douglas, J.C.; Loughran, R.M.; Rezendes, T.E.; Reed, K.R.; Cains, T.H.L.; Emsley, S.A.; Paddock, W.A.; Videau, P.; Koyack, M.J.; et al. Assessing the influence of curcumin in sex-specific oxidative stress, survival and behavior in *Drosophila melanogaster*. *J. Exp. Biol.* **2020**, *223*, jeb223867. [CrossRef]
29. Blackney, M.J.; Cox, R.; Shepherd, D.; Parker, J.D. Cloning and expression analysis of *Drosophila* extracellular Cu Zn superoxide dismutase. *Biosci. Rep.* **2014**, *34*, e00164. [CrossRef]
30. Phillips, J.P.; Campbell, S.D.; Michaud, D.; Charbonneau, M.; Hilliker, A.J. Null mutation of copper/zinc superoxide dismutase in *Drosophila* confers hypersensitivity to paraquat and reduced longevity. *Proc. Natl. Acad. Sci. USA* **1989**, *86*, 2761–2765. [CrossRef]
31. Celotto, A.M.; Liu, Z.; Vandemark, A.P.; Palladino, M.J. A novel *Drosophila* SOD2 mutant demonstrates a role for mitochondrial ROS in neurodevelopment and disease. *Brain Behav.* **2012**, *2*, 424–434. [CrossRef] [PubMed]
32. Shen, L.-R.; Parnell, L.D.; Ordovas, J.M.; Lai, C.-Q. Curcumin and aging. *BioFactors* **2013**, *39*, 133–140. [CrossRef] [PubMed]
33. Lee, K.-S.; Lee, B.-S.; Semnani, S.; Avanesian, A.; Um, C.-Y.; Jeon, H.-J.; Seong, K.-M.; Yu, K.; Min, K.-J.; Jafari, M. Curcumin extends life span, improves health span, and modulates the expression of age-associated aging genes in *Drosophila melanogaster*. *Rejuvenation Res.* **2010**, *13*, 561–570. [CrossRef] [PubMed]
34. Parks, A.L.; Cook, K.R.; Belvin, M.; Dompe, N.A.; Fawcett, R.; Huppert, K.; Tan, L.R.; Winter, C.G.; Bogart, K.P.; Deal, J.E.; et al. Systematic generation of high-resolution deletion coverage of the *Drosophila melanogaster* genome. *Nat. Genet.* **2004**, *36*, 288–292. [CrossRef]
35. Shishodia, S. Molecular mechanisms of curcumin action: Gene expression. *BioFactors* **2013**, *39*, 37–55. [CrossRef]
36. Ming, J.; Ye, J.; Zhang, Y.; Xu, Q.; Yang, X.; Shao, X.; Qiang, J.; Xu, P. Optimal dietary curcumin improved growth performance, and modulated innate immunity, antioxidant capacity and related genes expression of NF-κB and Nrf2 signaling pathways in grass carp (*Ctenopharyngodon idella*) after infection with *Aeromonas hydrophila*. *Fish Shellfish Immunol.* **2020**, *97*, 540–553. [CrossRef]
37. Shahcheraghi, S.H.; Salemi, F.; Peirovi, N.; Ayatollahi, J.; Alam, W.; Khan, H.; Saso, L. Nrf2 regulation by curcumin: Molecular aspects for therapeutic prospects. *Molecules* **2021**, *27*, 167. [CrossRef]
38. Zhu, H.; Itoh, K.; Yamamoto, M.; Zweier, J.L.; Li, Y. Role of Nrf2 signaling in regulation of antioxidants and phase 2 enzymes in cardiac fibroblasts: Protection against reactive oxygen and nitrogen species-induced cell injury. *FEBS Lett.* **2005**, *579*, 3029–3036. [CrossRef]
39. Ma, Q.; He, X. Molecular basis of electrophilic and oxidative defense: Promises and perils of Nrf2. *Pharmacol. Rev.* **2012**, *64*, 1055–1081. [CrossRef]
40. Signor, S.; Nuzhdin, S. Dynamic changes in gene expression and alternative splicing mediate the response to acute alcohol exposure in *Drosophila melanogaster*. *Heredity* **2018**, *121*, 342–360. [CrossRef]
41. McGraw, L.A.; Clark, A.G.; Wolfner, M.F. Post-mating gene expression profiles of female *Drosophila melanogaster* in response to time and to four male accessory gland proteins. *Genetics* **2008**, *179*, 1395–1408. [CrossRef] [PubMed]
42. Telonis-Scott, M.; van Heerwaarden, B.; Johnson, T.K.; Hoffmann, A.A.; Sgrò, C.M. New levels of transcriptome complexity at upper thermal limits in wild *Drosophila* revealed by exon expression analysis. *Genetics* **2013**, *195*, 809–830. [CrossRef] [PubMed]

BioTech

MDPI

Article

Phytochemical Composition, Antioxidant, Anti-*Helicobacter pylori*, and Enzyme Inhibitory Evaluations of *Cleistocalyx operculatus* Flower Bud and Leaf Fractions

Doan Thien Thanh [1,2], Mai Thanh Tan [1], Nguyen Thi My Thu [1], Pham Nhat Phuong Trinh [2], Pham Thi Hoai Thuong [2], Pham Thi Giang Tuyet [2], Luong Thi My Ngan [1,*] and Tran Trung Hieu [1,*]

[1] Faculty of Biology and Biotechnology, VNUHCM-University of Science, Ho Chi Minh City 700000, Vietnam; doanthienthanh@tdtu.edu.vn (D.T.T.); thanhtan.mtt1996@gmail.com (M.T.T.); nguyenmythu2112@gmail.com (N.T.M.T.)

[2] Faculty of Applied Sciences, Ton Duc Thang University, Ho Chi Minh City 700000, Vietnam; pnhatptrinh@gmail.com (P.N.P.T.); ptht.011001@gmail.com (P.T.H.T.); giangtuyet2605@gmail.com (P.T.G.T.)

* Correspondence: ltmngan@hcmus.edu.vn (L.T.M.N.); hieutt@hcmus.edu.vn (T.T.H.)

Abstract: Six solvent fractions isolated from flower bud and leaf ethanolic extracts of *Cleistocalyx operculatus* were analyzed for their phytochemical contents, including phenolics, flavonoids, saponins, tannins, and alkaloids. Antioxidant activities were measured using the ABTS, DPPH, and FRAP assays. The results showed that the flower bud aqueous fraction (BAF) and the leaf aqueous fraction (LAF) rich in phenolic content (768.18 and 490.74 mg GAE/g dry extract, respectively) exhibited significantly higher antioxidant activities than the other fractions. The flower bud hexane fraction (BHF) had remarkably high flavonoid and saponin contents (134.77 mg QE/g and 153.33 mg OA/g dry extract, respectively), followed by that of the leaf hexane fraction (LHF) (76.54 mg QE/g and 88.25 mg OA/g dry extract, respectively). The BHF and LHF were found to have extremely high antibacterial activity against two *H. pylori* strains, ATCC 51932 and 43504 (MICs of 125 µg/mL). Interestingly, DMC (2′,4′-Dihydroxy-6′-methoxy-3′,5′-dimethylchalcone) isolated from the BHF displayed greater antibacterial activity against the bacterial strains (MICs of 25–50 µg/mL) than those of the fractions. In addition, DMC presented potent inhibitory effects on *H. pylori* urease (IC_{50} of 3.2 µg/mL) and α-amylase (IC_{50} of 83.80 µg/mL), but no inhibition against α-glucosidase. It was also demonstrated that DMC showed pronounced inhibitory effects on the urease activity and biofilm formation of *H. pylori*, and could increase the membrane permeability of the bacterial cells. Scanning electron micrographs depicted that the BHF and DMC had strong effects on the cell shape and significantly induced the distortion and damage of the cell membrane. The fractions and DMC showed no significant toxicity to four tested human cell lines. Efforts to reduce antibiotic use indicate the need for further studies of the flower buds and DMC as potential products to prevent or treat gastric *H. pylori* infections.

Keywords: *Cleistocalyx operculatus*; antioxidant effect; anti-*H. pylori* activity; antibiofilm formation; membrane permeability; morphological transformation

Key Contribution: Flower bud hexane fraction (BHF) and leaf hexane fraction (LHF) of *Cleistocalyx operculatus* and isolated compound DMC possessed considerable activity against H. pylori. DMC had potential urease inhibitory and antibiofilm effects and increased the membrane permeability of the bacterial cells. SEM suggests that BHF and DMC caused strong effects on the cell shape and significantly damaged the cell membrane.

1. Introduction

The flower buds and leaves of *Cleistocalyx operculatus* (Roxb.) Merr. and L.M. Perry (or *Syzygium nervosum* DC.) are rich sources of bioactive compounds, predominately containing flavonoids, chalcones, and triterpenoids [1,2]. In Asian cultures, these herbal materials have been traditionally used as tea or remedies in the treatments of various disorders, such as influenza, bacillary dysentery, gastric inflammation, abdominal pain, and skin infection, as well as for their antiseptic properties [3,4]. Several solvent extracts of the flower buds have been known to exert various pharmacological activities in vitro and in vivo, including anti-hyperglycemic and cardio-tonic effects [5–8]. DMC (2′,4′-Dihydroxy-6′-methoxy-3′,5′-dimethylchalcone), a major constituent of the flower buds, was found to significantly inhibit the growth of human liver cancer cells and human umbilical vein endothelial cells [2,9]. Few works refer to the antibacterial activities of DMC, particularly towards *H. pylori*. DMC was reported to inhibit the growth of *Bacillus subtilis*, *Escherichia coli*, and *Cladosporium cucumerinum* [10]. A methanolic extract of the leaves was shown to have inhibitory activity against Gram-positive bacteria (e.g., *Staphylococcus aureus*, *Bacillus subtilis*, and *Streptococcus mutans*) and the yeast *Candida maltosa*, but no effect on Gram-negative bacteria (e.g., *Escherichia coli* and *Pseudomonas aeruginosa*) [6]. However, previous research indicated that the crude ethanolic extract of the leaves showed high antibacterial activity against *Helicobacter pylori* ATCC 51932 and three clinical isolates of *H. pylori* [11]. Recent research has reported that the crude hexane extract of *C. operculatus* flower buds exhibited the most growth-inhibitory activity against *Salmonella typhimurium* and *H. pylori*, while the crude ethanol and methanol extracts of the flower buds exhibited the strongest antioxidant activities [4]. Furthermore, the hexane extract was found to have the strongest inhibitory effect on *H. pylori* urease activity [11].

H. pylori is a Gram-negative bacterium with a curved or spiral shape that infects and colonizes the human gastric mucosa [12]. The bacterial infections are prevalent in over half of the world's population, particularly in developing regions, where the infections can affect as much as 90% of the population and tend to endure for a person's lifetime [13]. *H. pylori* infections cause asymptomatic chronic active gastritis in most infected individuals and lead to stomach ulcer disease and mucosa-associated lymphoid tissue lymphoma [14,15]. Several important factors, including vacuolating cytotoxin A, cytotoxin-associated gene A, chemotactic motility, adhesins, biofilm formation, and urease production of *H. pylori*, are known to contribute to the virulence of this organism [16–18]. Its spiral shape is the predominant form involved in the ability to thrive and colonize the gastric epithelial cells. Under unfavorable conditions, the spiral forms can convert into coccoid forms as a survival mechanism, but the coccoid forms are much less infective and virulent, and less likely to colonize and cause inflammation [18]. In addition, the biofilm formation of *H. pylori* provides the bacterium with protection and resistance to antimicrobial agents [19]. Nowadays, global antibiotic resistance in *H. pylori* is on the rise in many parts of the world, leading to treatment failure and reinfection [12]

This study aims to investigate the phytochemical contents in six solvent fractions of the *C. operculatus* flower buds and leaves, and evaluate their antioxidant, enzyme inhibitory, and anti-*H. pylori* activities. The effects of the fractions and DMC on biofilm formation, membrane permeability, and the cell morphology of *H. pylori* were also evaluated.

2. Materials and Methods

Reagents, chemicals, and enzymes utilized in the bioactivity studies were purchased from Sigma-Aldrich (Schnelldorf, Germany) and Merck (Damstadt, Germany). Two *Helicobacter pylori* strains (ATCC 51932 and ATCC 43504), MCF-7 cells (HTB-22, human breast cancer cell line), Jurkat cells (TIB-152, blood cancer cell line), and HeLa cells (CCL-2, cervical carcinoma cell line) were provided by ATCC (The American Type Culture Collection, Manassas, Rockville, MD, USA). Fibroblast cells derived from human foreskins were provided by the Laboratory of Molecular Biology, Department of Genetics, VNUHCM-University of Science [20]. Media and serum were obtained from HIMEDIA (Maharashtra, India), Becton, Dickinson and Company (Franklin Lakes, NJ, USA), Sparks (Aukland, New Zealand), Gibco-Thermo Fisher Sciemtific (Auckland, New Zealand), and Sigma-Aldrich (St. Louis,

MI, USA). All of the solvents and other chemicals were of reagent grade and commercially available.

2.1. General Experimental Procedures

An Agilent 1260 infinity series HPLC system (San Jose, CA, USA) with a ZORBAX Eclipse Plus C18 column (4.6 × 150 mm, 3.5 μm) was used, which was maintained at 25 °C with a UV detection at 320 nm. The mobile phase compositions were acetonitrile and 0.1% formic acid (70:30, v/v) with a flow rate of 1 mL/min. The injection volume was 10 μL with a running time of 75 min. The ^1H and ^{13}C NMR (Nuclear Magnetic Resonance) spectra were obtained in $CDCl_3$ at 25 °C using a 500 MHz Bruker AVANCE III HD spectrometer (Ettlingen, Germany). The Fourier transform infrared *spectroscopy* (FT-IR) spectrum was recorded by a Bruker Tensor 27 FT-IR Spectrometer (Bruker Optik GmbH, Ettlingen, Germany). Silica Gel 60 (0.06–0.2 mm) (Scharlab, Sentmenat, Spain) was used in a silica gel column (∅2 × 30 cm).

2.2. Extraction and Isolation of Plant Material

Dried flower buds and leaves of *C. operculatus* were collected from the traditional medicine and herbal material market on Hai Thuong Lan Ong Street, Dist. 5th, Ho Chi Minh City, Vietnam, and identified at the Department of Plant Biotechnology and Biotransformation, Faculty of Biology and Biotechnology, VNUHCM-University of Science, Ho Chi Minh City, Vietnam. Herbarium specimens of the flower buds and leaves were deposited in the Department under the code COFB1001 [4] and COL1002, respectively. Ethanolic crude extracts of the flower buds and leaves were prepared following the method described by Thanh et al. (2024) [4]. The extracts of the flower buds (90 g) and leaves (81 g) were partitioned by liquid–liquid fractionation to yield 1.8 g of flower bud hexane fraction (BHF), 78 g of flower bud ethyl acetate fraction (BEF), and 10.2 g of flower bud aqueous fraction (BAF), and 26.7 g of leaf hexane fraction (LHF), 24.3 g of leaf ethyl acetate fraction (LEF), and 30 g of leaf aqueous fraction (LAF), respectively, which are presented in the Supplementary Material (Figure S1).

Guided by the anti-*H. pylori* tests, the BHF (1.8 g), one of the most active fractions, was purified by silica gel column chromatography eluted with an *n*-hexane (H) and ethyl acetate (EA) (100:0–0:100) gradient to obtain six column fractions, fraction 1 (34.7 mg), fractions 2–3 (17.6 mg), and fractions 4–6 (56.4 mg), described previously by [9] (Figure S1). Repetitive column chromatography of fractions 2–3 (17.6 mg) eluted with H/AE = 90:10 (Figure S1) resulted in an active principle 1 (14.2 mg) as yellow needle-shaped crystals.

The purity of principle 1 was found to be greater than 95% by the HPLC analysis (Figure S2), and its spectroscopic data matched the data reported in previous research by Choommongkol et al. (2022) [21]. The principle 1, DMC (2′,4′-Dihydroxy-6′-methoxy-3′,5′-dimethylchalcone), was identified based on the following evidence: FT-IR ν (KBr): the stretching vibrations of the O-H (3421 cm^{-1}), $\nu_{Ar\text{-}H}$ and $\nu_{=C\text{–}H}$ (3001–3028 cm^{-1}), the stretching vibrations of the C–H (2855–2925 cm^{-1}), the C=O (1626 cm^{-1}), the C_{Ar}=C_{Ar} (1451–1540 cm^{-1}), the deformation vibrations of the C-H (1359–1419 cm^{-1}), the stretching vibrations of the C-O (1111–1220 cm^{-1}), and the out-of-plane deformation vibrations of the C_{Ar}-H and =C–H (612–987 cm^{-1}) (Figure S3). The ^1H NMR, ^{13}C NMR, and HMBC data of DMC are presented in Table S1 and Figures S4 and S5. The spectral data were identical to the published data of Choommongkol et al. (2022) [21]. Additionally, the ^1H–^{13}C HMBC-correlations of DMC are presented in Figure S6.

2.3. Total Phytochemical Contents

The total phenolic content (TPC) was quantified using the Folin–Ciocalteu assay, as indicated by Temesgen et al. (2022) [22]. An aliquot of each fraction (0.9 mL) and 4.5 mL of Folin–Ciocalteu reagent (10%) was vigorously shaken and incubated for 5 min in the dark. Then, 1.8 mL of Na_2CO_3 (7%) was added into the mixture and continued to incubate for 30 min. Absolute ethanol was used as a control (blank), and gallic acid (0–100 μg/mL) was also prepared to build a standard curve (y = 0.0095x + 0.0029; R^2 = 0.9978). Absorbance at 765 nm was subsequently measured using a spectrophotometer (UV-5100, Metash, Shanghai, China). The TPC was conveyed as the mg gallic acid equivalent (GAE)/g of dry fraction.

The total flavonoid content (TFC) was determined following the aluminum chloride colorimetric method [23]. An aliquot of each fraction (300 μL) was mixed with 150 μL of $NaNO_2$ solution (5%) and incubated for 6 min at room temperature. The mixture was then added with 300 μL of $AlCl_3$ (5%) and 1 mL of NaOH 1 M and allowed to stand for 6 min. Absolute ethanol was prepared as a control (blank) to replace the extract. Quercetin (0–100 μg/mL) was used to estimate the standard curve (y = 0.0075x − 0.0017; R^2 = 0.9998). The absorbance of the mixture was measured at 510 nm. The TFC was stated as the mg of quercetin equivalent (QE)/g of dry fraction.

The total alkaloid content (TAC) was quantified following the modified method of Ncube et al. (2015) [24]. Briefly, 1 mL of each fraction diluted in HCl (2N) was mixed well with 5 mL of bromocresol green (BCG) solution (0.01%) and 5 mL of phosphate buffer solution at a pH of 4.7 in a separating funnel. The mixture was vigorously shaken and extracted twice with 5 mL of chloroform. Then, the collected alkaloid extracts were diluted with 10 mL of chloroform. The absorbance of the mixture was read at 470 nm. Atropine (0–100 μg/mL) and absolute ethanol were used as the standard solution (y = 0.0128x − 0.0493; R^2 = 0.9927) and control solution, respectively. The total content of alkaloids was expressed as the mg atropine equivalent (AE)/g of dry fraction.

The total tannin content (TTC) was quantified using a previously reported method [23]. An aliquot of each fraction (0.5 mL) was added into 3.0 mL of vanillin solution (4% in methanol, *w*/*v*). Then, the mixture was stirred with 1.5 mL of HCl and incubated for 15 min in the dark. The absorbance of the mixture was measured at 500 nm, and absolute ethanol was the control (blank) to replace the extract. The TTC was calculated using the catechin standard curve (y = 0.0012x + 0.0261; R^2 = 0.9984) and shown as the mg of catechin equivalent (CE)/g of dry fraction.

The total saponin content (TSC) was determined using the vanillin method [25]. Briefly, each fraction (0.5 mL) was mixed with 0.1 mL of vanillin solution (5% in acetic acid, *w*/*v*) and 0.4 mL of 70% perchloric acid. The mixture was heated at 60 °C for 15 min in a water bath and then cooled to room temperature. After that, 5 mL of absolute acetic acid was mixed well into the mixture. Oleanolic acid (0–300 μg/mL) was used to measure the standard curve (y = 0.0041x − 0.0256; R^2 = 0.9934). The absorbance values were measured at 548 nm. The TSC was expressed as the mg oleanolic acid equivalent (OAE)/g of dry fraction.

2.4. Antioxidant Assay

A DPPH (2,2-diphenyl-1-(2,4,6-trinitrophenyl)hydrazyl) radical scavenging assay was followed according to a previous study, as described by Elouafy et al. (2023) [26], to assess the antioxidant activity. Briefly, different concentrations of each fraction (0–150 μg/mL) were prepared in methanol. Each dilution (1 mL) was well mixed with 1 mL of DPPH methanolic solution (25 μg/mL) and incubated at 37 °C in the dark for 30 min. The absorbance of the mixture was subsequently measured at 517 nm. Ascorbic acid and absolute methanol were used as the positive control and blank, respectively. The DPPH scavenging activity was measured using Equation (1), as follows:

$$\text{DPPH scavenging activity } (\%) = \frac{OD_1 - OD_2}{OD_1} \times 100\ (\%) \tag{1}$$

where OD_1 and OD_2 are the absorbance values of the blank and each fraction or ascorbic acid, respectively. The concentration required to inhibit 50% of the free DPPH radial (IC_{50}) was calculated by plotting the DPPH scavenging activity versus the sample concentration.

The measurement of the antioxidant activity was also performed using the ABTS•+ (2,2′-azinobis(3-ethylbenzothiazoline-6-sulfonic acid) diammonium salt) assay presented by Olszowy-Tomczyk and Typek (2024) [27], with some modifications. Briefly, the initial solutions consisted of 7 mM ABTS•+ solution (5 mL) and 140 mM $K_2S_2O_8$ solution (88 μL). The working solution was reacted in the dark at room temperature for 16 h and then diluted in ethanol until the absorbance value of 0.7 ± 0.002 at 734 nm using a spectrophotometer. Fresh ABTS•+ solution was created for each assay. For the analysis, each fraction (0.01 mL) was mixed with 0.99 mL of the ABTS•+ solution. The absorbance of the mixture was then measured at 734 nm after 6 min of storage in the dark. After 6 min of storage in the dark, the absorbance was measured at 734 nm. Absolute methanol and Trolox (6-hydroxy-2,5,7,8-tetramethylchroman-2-carboxylic acid) were used as the negative control and positive control, respectively. The ABTS•+ scavenging capacity was evaluated as the percentage of inhibition of the ABTS radical scavenging activity using Equation (2), as follows:

$$\text{ABTS radical scavenging activity } (\%) = \frac{OD_1 - OD_2}{OD_1} \times 100\% \quad (2)$$

where OD_1 and OD_2 are the absorbance of the ABTS•+ radical cation in methanol and in each fraction or Trolox, respectively. The concentration required to inhibit the 50% ABTS•+ solution (IC_{50}) was subsequently estimated by plotting the ABTS radical scavenging activity versus the sample concentration.

The FRAP assay was established using a modified method [23]. Briefly, the FRAP reagent was prepared by mixing 300 mM sodium acetate buffer at a pH of 3.6 (25 mL), 10 mM TPTZ (2, 4, 6- tripyridyl-s-triazine) solution in 40 mM HCl (2.5 mL), and 20 mM $FeCl_3.6H_2O$ solution (2.5 mL), and then warmed at 37 °C before use. An amount of 150 μL of each fraction was mixed with 2850 μL of the FRAP reagent and then incubated at 37 °C for 30 min in the dark in order to initiate the reaction. The standard curve of the assay was also prepared by using a serial concentration of Trolox (0–500 μg/mL). The absorbance of the colored product was read at 593 nm. The antioxidant capacity was stated as the mg Trolox equivalent (TE)/g of the dry fraction.

2.5. *Anti-Helicobacter pylori Assay*

Two *H. pylori* strains were stored in BHI (brain heart infusion) broth medium supplemented with 10% NBS (newborn bovine serum) containing vancomycin (10 mg/L), polymyxin B (5 mg/L), trimethoprim (5 mg/L), and amphotericin B (2 mg/L), and 25% glycerol in a liquid nitrogen container until use. The minimal inhibitory concentrations (MICs) of the fractions and DMC were established using a broth dilution assay in sterile 96-well plates for the bacterial strains [28]. Amount of 25 μL of the tested samples at different concentrations from 1 to 1000 μg/mLdiluted in DMSO was added to 75 μL of Brucella broth supplemented with 10% NBS.The final concentration of the DMSO in the assay was less than 2.5%. Generally, 30 μL of bacterial suspension (5 × 10^6 CFU/mL) of each strain was added to 10 μL of tested samples. Then, the plates were incubated at 37 °C and shaken at 50 rpm for 48 h in a microaerophilic condition using a Oxoid CampyGen sachet (Thermo Fisher Scientific, Loughborough, UK) placed in a sealed jar. The MIC value was defined as the lowest concentration that visibly inhibited bacterial growth using resazurin as an indicator.

2.6. *Enzyme Inhibitory Assay*

Crude urease of *H. pylori* ATCC 43504 was prepared according to the method of Ngan et al. (2012) [29]. Briefly, 10 μL of each fraction or DMC at various concentrations of 0–500 μg/mL was added in 30 μL of 20 mM EDTA–sodium phosphate buffer (pH of 7.3). Then, 10 μL of urease solution (75 μL of urease/1 mL of the buffer) was added and incubated at room temperature for 1 h before adding 50 μL of urea solution (0.24 mg of urea/1 mL of the buffer), and allowed to incubate at room temperature for 30 min. After that, 40 μL of solution A (40% sodium salicylate and 0.3% sodium nitroprusside) and 60 μL of solution B (0.5% sodium hydroxide and 0.042% sodium hypochlorite) were added into the mixture. The ammonia generated by the urease activity was quantified by measuring the absorbance on a Microlisa Plus microplate reader (Micro Lab Instruments, Ahmedabad, India) at 625 nm with ammonium chloride as a standard and buffer solution as a control.

The α-glucosidase activity was determined by measuring the release of p-nitrophenol from pNPG (p-nitrophenyl-α-D-glucopyranoside) according to the method of Shai et al. (2011) [30], with slightly modification. A serial concentration (0–500 μg/mL) of each fraction and DMC were dissolved in DMSO (dimethyl sulfoxide). A mixture containing 40 μL of tested samples, 20 μL of α-glucosidase (2.0 U/mL), and 100 μL of potassium phosphate buffer (100 mM, pH = 6.8) was prepared and preincubated at 37 °C for 15 min. The reaction was initiated by adding 40 μL of pNPG 5 mM and incubated at 37 °C for 20 min. To stop the reaction, 100 μL of Na_2CO_3 0.1 M was added to the solution. The amount of p-nitrophenol released by α-glucosidase was defined as measuring the absorbance at 405 nm. Acarbose was used as the standard inhibitor. The control solution contained buffer solution (40 μL) instead of the fractions, DMC, or standard inhibitor.

The α-amylase inhibitory activity of the fractions and DMC were caried out according to the method of Ogunyemi et al. (2022) [31], with slightly modifications. A mixture of 100 μL of various concentrations (0–500 μg/mL in DMSO) of each fraction or DMC and 100 μL of sodium phosphate buffer (0.02 M, pH of 6.9) containing 200 μL of α-amylase solution (2.5 U/mL) were incubated at room temperature for 10 min. After preincubation, the mixture was added with 200 μL of 1% soluble starch solution in sodium phosphate buffer. The mixture was reacted at 30 °C for 10 min before adding of 300 μL of HCl 1 N to stop the reaction. The mixture was added to 90 μL of 10% iodine solution to observe the color change and diluted with 10 mL of distilled water. The sodium phosphate buffer and acarbose served as the control and standard inhibitor, respectively. The absorbance was read at 540 nm.

These enzyme inhibitory activities were determined using Equation (3), as follows:

$$\text{Inhibition activity}(\%) = \frac{(\text{Ac} - \text{As})}{\text{Ac}} \times 100\% \tag{3}$$

where Ac is the absorbance of the control and As is the absorbance of the fractions, DMC, or standard inhibitors.

2.7. Biofilm Formation Inhibitory Assay

The antibiofilm activity of the fractions and DMC was carried out in 96-well plates following the method of Hieu et al. (2022) [32]. In brief, 30 μL of *H. pylori* ATCC 43504 suspension (10^8 CFU/mL) and 10 μL of the tested samples at sub-MICs (MIC/2, MIC/4, and MIC/8) were added into 60 μL of Brucella broth in each well of sterile 96-well plates. Blank wells containing DMSO and background wells containing samples were also similarly arranged as control wells without the bacterial suspension. After 48 h of incubation at 37 °C in a microaerobic environment with shaking at 150 rpm, the medium was removed by using PBS (0.01 M phosphate-buffered saline, pH of 7.2) to wash the wells. The plates were then air-dried before being fixed with absolute methanol for 15 min and dried. Each well was dyed with a 0.1% crystal violet solution for 10 min and washed with distilled water. The dried plates were dissolved with crystal violet using 95% ethanol for 15 min. The Microlisa Plus microplate reader was used to measure the absorbance at 595 nm.

2.8. Scanning Electron Microscopy

To determine the efficacy of the BHF and DMC against *H. pylori* through the morphological changes, a scanning electron microscopy (SEM) analysis was performed [33]. Overnight broth cultures of *H. pylori* ATCC43504 were prepared in Brucella broth. The cell suspension of *H. pylori* was cultured after 48 h with or without tested samples at MIC concentrations. Then, the bacteria were harvested by centrifugation at 14,000 rpm for 5 min. Briefly, the specimens were fixed in modified Karnovsky's fixative (2% glutaraldehyde and 2% paraformaldehyde in 0.05 M sodium cacodylate buffer, pH of 7.2) for 4 h, and then post-fixed in 1% osmium tetroxide in 0.05 M sodium cacodylate buffer for 4 h at 4 °C.

A series of increasing concentrations of ethanol (20, 40, 60, 80, 95%, and absolute) and hexamethyldisilazane were used to dehydrate the specimens, each twice for 15 min. SEM was performed by the Institute of Chemical Technology, Ho Chi Minh City, Vietnam. The specimens were then mounted on SEM stubs by double-sided carbon conductive tape and coated with gold (JEC-3000FC ion sputter, JEOL Co., Tokyo, Japan). The SEM images were obtained using a scanning electron microscope, the JSM-IT200 (JEOL Co., Japan), operating at an acceleration voltage of 5 kV.

2.9. Membrane Permeability Assay

The membrane permeability of *H. pylori* was determined by a crystal violet assay [34]. Briefly, 30 µL suspensions of *H. pylori* ATCC 43504 (10^8 CFU/mL) were prepared in 60 µL of BB medium and 10 µL of each fraction and DMC at different concentrations (MIC, MIC/2, MIC/4, and MIC/8). The mixture was incubated for 2 h and the cells were harvested at 4500 rpm for 5 min at 4 °C. The cells were washed twice in PBS (0.01 M, pH of 7.2). After that, the cells were resuspended in 1 mL of PBS containing 100 µL of 0.1% crystal violet and incubated for 5 min at 37 °C. The suspension was then centrifuged at 14,000 rpm for 15 min and the supernatant was obtained with the Microlisa Plus microplate reader at 590 nm. The percentage of the crystal violet uptake of the *H. pylori* cells in all of the tested samples was calculated using Equation (4), as follows:

$$\text{Crystal violet uptake}(\%) = \frac{(1 - \text{OD value of the sample})}{\text{OD value of crystal violet solution}} \times 100\% \quad (4)$$

2.10. Cell Line Cultures and Cytotoxicity Assay

MCF-7, HeLa, Jurkat, and fibroblast cell lines were grown in EMEM (Eagle's minimal essential medium) for the MCF-7 and HeLa cells, RPMI (*Roswell Park Memorial Institute*) medium for the Jurkat cells, and DMEM/F12 (Dulbecco's modified eagle medium: nutrient mixture F12) medium for the fibroblast cells. These media were supplemented with 10% FBS (fetal bovine serum), 2 mM L-glutamine, 20 mM HEPES, 0.025 µg/mL of amphotericin B, 100 IU/mL of penicillin G, and 100 µg/mL of streptomycin at 37 °C and 5% CO_2. The 2nd to 5th passages of the cultivations of the fibroblast cells, and the 4th to 20th passages of the cultivations of the HeLa, MCF7, and Jurkat cells, were used.

The SRB (Sulforhodamine B) assay was performed according to Nguyen and Huynh (2016) [20]. In brief, the cells were seeded in 96-well plates at a density of 10,000 cells/well for the MCF-7, HeLa, and fibroblast cells, and 50,000 cells/well for the Jurkat cells. These cells were then cultured for 24 h before exposure to varying concentrations of each fraction or DMC for a duration of 48 h. Treated cells were fixed with a cold 50% (*w*/*v*) trichloroacetic acid solution for 1–3 h, then washed and stained with 0.2% (*w*/*v*) SRB for 20 min. After washing with 1% acetic acid five times, protein-bound dye was solubilized in a 10 mM Tris base solution. Optical density values were measured using the Microlisa Plus microplate reader at wavelengths of 492 nm and 620 nm. Camptothecin and 0.25% DMSO were used as the positive and negative controls, respectively. The percentage of growth inhibition was determined using Equation (5), as follows:

$$\text{Inhibition}(\%) = 1 - \frac{\text{ODt}}{\text{ODc}} \times 100\% \quad (5)$$

where ODt is the optical density value of the tested sample and ODc is the value of the control sample.

2.11. Statistical Analysis

The MIC values of each test fraction and compound were established with at least three independent experiments performed in triplicate ($n \geq 9$). Tested materials with MIC values of ≤130, >130–<630, 630–1250, >1250–<2500, and ≥2500 µg/mL were classified as extremely high, high, moderate, low, and no inhibitory activity against bacteria growth, respectively [35]. All other experiments were performed in triplicate and the data are shown as mean ± standard derivations (SD) (n ≥ 3). Analysis of variance (ANOVA) by Tukey's multiple comparison, tested as $p < 0.05$, and the half-maximal inhibitory concentration value (IC_{50}) and half-cytotoxicity concentration (CC_{50}) value were conducted using the GraphPad Prism 8 software program (San Diego, CA, USA).

3. Results

3.1. Total Phytochemical Contents

The total phytochemical contents of different fractions from the flower buds and leaves of *C. operculatus* are shown in Table 1. The BAF and LAF revealed the highest values of the TPC, with 768.18 and 490.74 mg GAE/g dry extract, respectively. The lowest TPC was found in the LHF (201.63 mg GAE/g dry extract). In contrast, the BHF showed the highest TFC and TSC (134.77 mg QE/g dry extract and 153.33 mg OA/g dry extract, respectively), followed by the BEF (85.88 mg QE/g dry extract and 158.10 mg OA/g dry extract, respectively) and the LHF and LEF (71.72–76.54 mg QE/g dry extract and 75.24–88.25 mg OA/g dry extract, respectively). The LAF contained the lowest levels of both the TFC and TSC (25.56 mg QE/g dry extract and 25.56 mg OA/g dry extract,) while the BAF had the lowest level of the TFC (11.04 mg QE/g dry extract). In the experiment, the flower bud and leaf fractions of *C. operculatus* presented relatively low values of the TTC (4.36–42,97 mg CE/g dry extract) and significantly low values of the TAC (1.50–5.41 mg AE/g dry extract).

Table 1. Total phytochemical contents in different fractions from flower buds and leaves of *C. operculatus*.

Fractions	TPC	TFC	TSC	TTC	TAC
BHF	426.77 ^{c} ± 1.22	134.77 ^{a} ± 7.75	153.33 ^{a} ± 4.69	42.97 ^{a} ± 2.93	1.66 ^{d} ± 0.09
BEF	280.46 ^{d} ± 11.06	85.88 ^{b} ± 2.52	158.10 ^{a} ± 5.97	22.97 ^{c} ± 1.73	1.50 ^{d} ± 0.17
BAF	768.18 ^{a} ± 12.20	11.04 ^{e} ± 0.53	81.59 ^{b} ± 1.20	13.53 ^{d} ± 2.93	3.04 ^{c} ± 0.20
LHF	201.63 ^{f} ± 3.45	76.54 ^{bc} ± 1.72	88.25 ^{b} ± 9.46	33.17 ^{b} ± 2.10	4.81 ^{b} ± 0.09
LEF	238.47 ^{e} ± 4.18	71.72 ^{c} ± 1.74	75.24 ^{b} ± 5.95	10.39 ^{d} ± 1.29	4.73 ^{b} ± 0.05
LAF	490.74 ^{b} ± 7.29	25.56 ^{d} ± 0.21	25.56 ^{c} ± 2.44	4.36 ^{e} ± 0.96	5.41 ^{a} ± 0.08

BHF: flower bud hexane fraction, BEF: flower bud ethyl acetate fraction, and BAF: flower bud aqueous fraction. LHF: leaf hexane fraction, LEF: leaf ethyl acetate fraction, and LAF: leaf aqueous fraction. TPC: total phenolic content (mg GAE/g dry extract), TFC: total flavonoid content (mg QE/g dry extract), TSC: total saponin content (mg OA/g dry extract), TTC: total tannin content (mg CE/g dry extract), and TAC: total alkaloid content (mg AE/g dry extract). Data are shown as mean ± SD (n = 3). Means within a column with different letters indicate a significant difference at $p < 0.05$.

3.2. Antioxidant Activity

The antioxidant activities of the different fractions from the flower buds and leaves of *C. operculatus* are exhibited in Table 2. Depending on the solvent used for extracting the fractions, the antioxidant activity of the fractions significantly varied. In the FRAP assay, the LAF exhibited the highest antioxidant activity (301.82 mg TE/g dry extract), followed by the BAF (201.80 mg TE/g extract). The lowest FRAP activity was obtained with the LHF (55.57 mg TE/g extract) and BHF (69.01 mg TE/g extract). The result of the DPPH and ABTS radical scavenging abilities indicated that the LAF exhibited the strongest antioxidant activity (IC_{50} values of 11.24 and 0.55 μg/mL, respectively) followed by the LEF (16.05 and 0.98 μg/mL) and BAF (24.69 and 1.08 μg/mL). However, the result from the ABTS method showed that the LEF, LHF, and BAF provided similar ABTS radical scavenging activity (0.98–1.08 μg/mL). Weak DPPH radical scavenging activity was found in the BHF, LHF, and BEF (IC_{50} range of 33.99–58.46 μg/mL), while weak ABTS radical scavenging activity was found in the BEF and BHF (1.48–1.70 μg/mL). All of the tested fractions produced lower antioxidant effects than that of ascorbic acid (IC_{50} of 3.34 μg/mL) but higher than that of the Trolox (IC_{50} of 2.63 μg/mL).

Table 2. Antioxidant activity of the different fractions from *C. operculatus* flower buds and leaves

Samples	FRAP (mg TE/g Extract)	DPPH IC_{50} (μg/mL)	ABTS IC_{50} (μg/mL)
BHF	69.01 [e] ± 1.833	33.99 [c] ± 0.76	1.70 [b] ± 0.09
BEF	91.95 [c] ± 1.302	58.46 [a] ± 1.370	1.48 [b] ± 0.162
BAF	201.80 [b] ± 4.502	24.69 [d] ± 0.194	1.08 [c] ± 0.013
LHF	55.57 [f] ± 1.265	39.96 [b] ± 0.237	1.06 [c] ± 0.084
LEF	78.56 [d] ± 1.13	16.05 [e] ± 0.031	0.98 [c] ± 0.007
LAF	301.82 [a] ± 2.306	11.24 [f] ±1.524	0.55 [d] ± 0.004
Ascorbic acid	ND	3.34 [g] ± 0.017	ND
Trolox	ND	ND	2.63 [a] ± 0.05

BHF: flower bud hexane fraction, BEF: flower bud ethyl acetate fraction, and BAF: flower bud aqueous fraction. LHF: leaf hexane fraction, LEF: leaf ethyl acetate fraction, and LAF: leaf aqueous fraction. Data are shown as mean ± SD (n = 3). Means within a column with different letters indicate a significant difference at $p < 0.05$. ND: not determined.

3.3. Anti-H. pylori Activity

The bacterial growth inhibitory activities of the *C. operculatus* fractions and DMC against two *H. pylori* strains (ATCC 51932 and ATCC 43504) are shown in Table 3. Both the BHF and LHF showed extremely high growth inhibitory activity against the both strains, with an MIC value of 125 μg/mL. Other fractions displayed a high growth inhibitory effect, with MIC values ranging from 250 to 500 μg/mL on both bacterial strains, except the LAF, with a moderate effect (an MIC value of 1000 μg/mL) on the strain 43504. Remarkably, DMC revealed stronger antibacterial activity (MIC values of 25 and 50 μg/mL) than those of the BHF and LHF. The amoxicillin MIC value against the strains was 0.01 μg/mL. This indicated that *H. pylori* ATCC 51932 and ATCC 43504 were susceptible amoxicillin strains, as the MIC resistance breakpoints for amoxicillin against *H. pylori* was reported to be >0.5 μg/mL [35].

Table 3. In vitro minimal inhibitory concentration (MIC) values of *C. operculatus* fractions and DMC against *H. pylori*.

Samples	MIC (μg/mL)	
	H. pylori ATCC 51932	*H. pylori* ATCC 43504
BHF	125	125
BEF	250	500
BAF	500	500
LHF	125	125
LEF	250	500
LAF	500	1000
DMC	25	50
Amoxicillin	0.01	0.01

BHF: flower bud hexane fraction, BEF: flower bud ethyl acetate fraction, and BAF: flower bud aqueous fraction. LHF: leaf hexane fraction, LEF: leaf ethyl acetate fraction, and LAF: leaf aqueous fraction. Test materials with MIC values of ≤130, >130–<630, 630–1250, >1250–<2500, and ≥2500 μg/mL were classified as the extremely high, high, moderate, low, and no inhibitory activity against the growth of test bacteria, respectively [36].

3.4. Enzyme Inhibitory Activity

The inhibitory activities of the *C. operculatus* fractions and DMC against *H. pylori* urease, α-glucosidase, and α-amylase are presented in Table 4. All of the tested fractions and DMC exhibited strong inhibition against *H. pylori* urease (IC_{50} values ranging from 2.3 to 3.6 μg/mL), except for the BEF (IC_{50} of 4.9 μg/mL) and LAF (IC_{50} of 6.8 μg/mL). These fractions and DMC showed better inhibitory activity against urease than that of thiourea (IC_{50} of 44.3 μg/mL) as a positive control. These fractions also had a strong inhibitory effect on α-glucosidase (IC_{50} values ranging from 0.6 to 2.6 μg/mL) and significantly stronger than the positive control, acarbose (IC_{50} of 25.6 μg/mL), whereas DMC had no inhibition against α-glucosidase (IC_{50} of 94.6 μg/mL). Nevertheless, these fractions exhibited weak (IC_{50} values ranging from 191.3 to 497.2 μg/mL) and no (IC_{50} of 1281.7 μg/mL) inhibitory activities against α-amylase compared with acarbose and DMC, which had similar inhibitory effects (IC_{50} values of 75.0 and 83.8 μg/mL, respectively).

Table 4. Inhibitory effects of *C. operculatus* fractions and DMC on *H. pylori* urease, α-glucosidase, and α-amylase.

Samples	IC_{50} (μg/mL)		
	H. pylori Urease	α-Glucosidase	α-Amylase
BHF	$2.3^{e} \pm 0.13$	$1.5^{c} \pm 0.09$	$398.5^{d} \pm 5.3$
BEF	$4.9^{c} \pm 0.40$	$0.9^{c} \pm 0.01$	$497.2^{b} \pm 17.1$
BAF	$2.5^{de} \pm 0.21$	$0.8^{c} \pm 0.01$	$444.3^{c} \pm 14.3$
LHF	$3.2^{de} \pm 0.26$	$2.6^{c} \pm 0.16$	$191.3^{f} \pm 9.5$
LEF	$3.6^{d} \pm 0.01$	$1.2^{c} \pm 0.06$	$292.6^{e} \pm 16.6$
LAF	$6.8^{b} \pm 0.07$	$0.6^{c} \pm 0.02$	$1281.7^{a} \pm 23.7$
DMC	$3.2^{de} \pm 0.03$	$94.6^{a} \pm 2.57$	$83.80^{g} \pm 0.08$
Thiourea	$44.3^{a} \pm 1.12$	ND	ND
Acarbose	ND	$25.6^{b} \pm 0.70$	$75.0^{g} \pm 3.86$

BHF: flower bud hexane fraction, BEF: flower bud ethyl acetate fraction, and BAF: flower bud aqueous fraction. LHF: leaf hexane fraction, LEF: leaf ethyl acetate fraction, and LAF: leaf aqueous fraction. Data are shown as mean ± SD (n = 3). Means within a column with different letters indicate a significant difference at $p < 0.05$. ND: not determined.

3.5. Effect on Biofilm Formation of *H. pylori*

The results of the biofilm inhibition revealed that all of the tested fractions and DMC at sub-MICs inhibited the biofilm formation of *H. pylori* after 48 h of treatments (Figure 1). At MIC/2, all of the fractions (62.5–500 μg/mL) and DMC (25 μg/mL) demonstrated potential antibiofilm activity and reduced biofilm formation by 82.2 ± 0.74–93.6 ± 1.53%, with a nonsignificant difference ($p > 0.05$). At MIC/4, the LEF (125 μg/mL) and DMC (12.5 μg/mL) induced pronounced antibiofilm effects, with inhibitory percentages of 78.4 ± 0.98 and 77.8 ± 1.37%, respectively, followed by the slightly lower activity of the BHF (31.25 μg/mL) and BEF (125 μg/mL), reducing the biofilm formation by 73.5 ± 0.98% and 70.2 ± 2.55%, respectively. However, the LHF (31.25 μg/mL), BAF (125 μg/mL), and LAF (250 μg/mL) reduced biofilm formation by 50.9 ± 6.86%, 54.3 ± 2.43%, and 57.8 ± 0.88%, respectively. Notably, at MIC/8, the BEF and LEF (62.5 μg/mL) remained able to inhibit biofilm formation by 58.2 ± 0.59% and 57.1 ± 0.98%, respectively, followed by DMC (6.25 μg/mL), which inhibited biofilm formation by 46.6% ± 0.49%. Other fractions had weak inhibitory effects (21.1 ± 0.49–36.4 ± 2.84%)

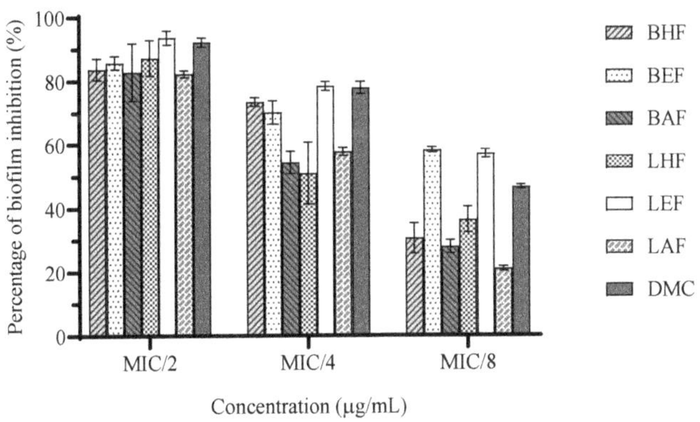

Figure 1. Effect of *C. operculatus* fractions and DMC at sub-MICs on *H. pylori* biofilm formation 48 h post-treatment. Data are reported as means ± SD.

3.6. Effect on the Morphology of H. pylori

The scanning electron micrographs for the untreated *H. pylori* showed that the cells appeared mainly in a spiral-shaped form that successfully incorporated within the biofilms (Figure 2A,B). The cell surface was smooth and regular with an intact cell membrane. However, after 48 h of treatment with the BHF at an MIC = 125 μg/mL (Figure 2C,D) and DMC at an MIC = 50 μg/mL (Figure 2E,F), 86.2 ± 3.56 and 93.7 ± 3.68% of the spiral cells were converted into coccoid-shaped cells, respectively. These figures indicated that the BHF and DMC could inhibit bacterial biofilm formation and induce clusters of coccoid cells. The clusters were aggregated and stuck to each other with rough or deformed membrane surfaces and cell debris. This suggests that the majority of coccoid forms induced by the BHF and DMC revealed the morphological manifestations of bacterial cell death or destruction.

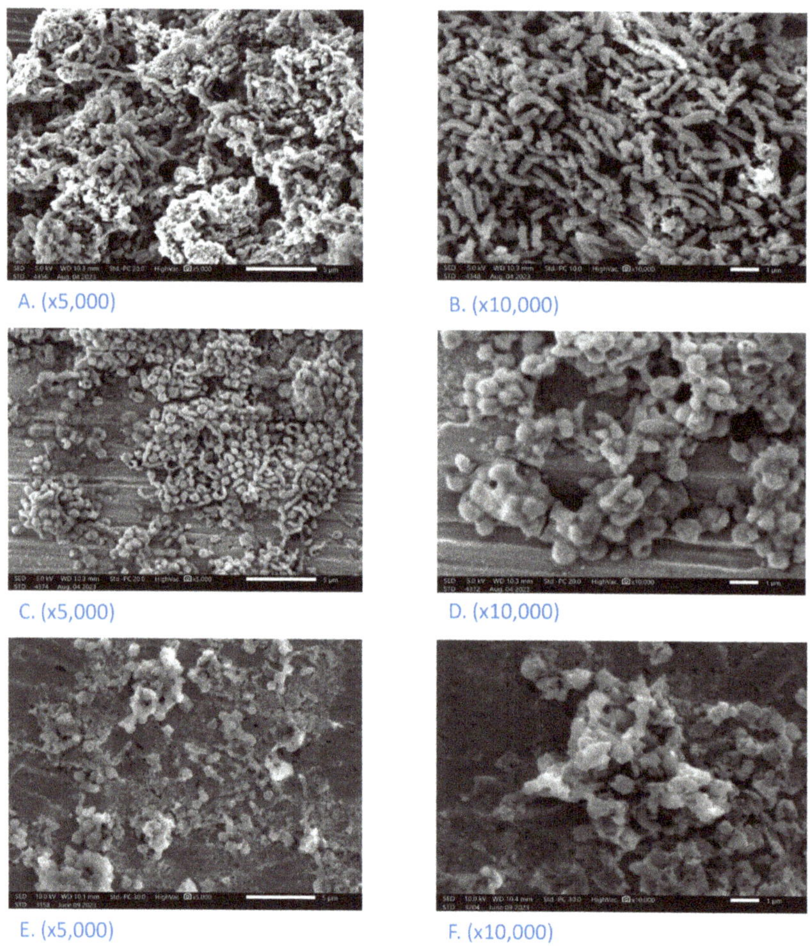

Figure 2. SEM micrographs of *H. pylori* ATCC 43504 depicting untreated cells (**A**,**B**) and cells treated with 125 μg/mL of BHF (**C**,**D**) and 50 μg/mL of DMC (**E**,**F**); bars 5 and 1 μm, respectively.

3.7. Effect on Membrane Permeability

The alteration in the membrane permeability of *H. pylori* caused by *C. operculatus* fractions and DMC is presented in Figure 3. The crystal violet uptake by untreated cells was only by 13.8 ± 1.04–26.0 ± 0.98%, but the crystal violet uptake by treated cells increased from 63.0 ± 0.87 to 71.3 ± 1.46%, which was caused by all of the tested fractions, and 79.8 ± 1.06%, which was caused by DMC at MIC/2 after 2h of treatment. At MIC/4, DMC (12.5 µg/mL) induced an extreme increase in the crystal violet uptake by 64.4 ± 2.76%, followed by those of the LHF and BHF (31.25 µg/mL), LEF and BAF (125 µg/mL), and LAF (250 µg/mL), increasing by 49.5 ± 2.43% and 50.3 ± 1.54%, 46.3 ± 4.17% and 54.8 ± 2.49%, and 50.1 ± 1.36%, respectively, and with the BEF (125 µg/mL), increasing by only 33.82 ± 1.73%. At MIC/8, the membrane permeability of the bacterial cells did not alter much (27.4 ± 1.90–40.4 ± 2.06%) compared with those of the untreated groups.

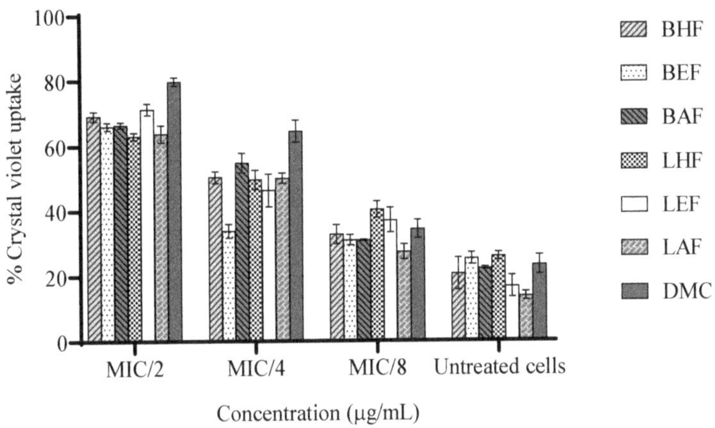

Figure 3. Effect of *C. operculatus* fractions and DMC at sub-MICs on the uptake of crystal violet by *H. pylori* ATCC 43504 after 2 h of treatment. Data are reported as means ± SD (n = 3).

3.8. Cytotoxicity Effects

In order to investigate the selectivity of the anti-*H. pylori* activity of *C. operculatus* fractions and DMC, the cytotoxic activity of these tested samples on four human cell lines was examined, as shown in Table 5. All of the fractions and DMC had no significant cytotoxicity against fibroblast cells, with CC_{50} values of >100 µg/mL. Among these factions, the BAF exhibited no cytotoxic activity against the four tested cell lines (CC_{50} values > 100 µg/mL), whereas the BHF displayed the most cytotoxic effect on three cancer cell lines, Jurkat, MCF-7, and HeLa cells, with CC_{50} values of 18.51, 30.79, and 31.70 µg/mL, respectively. Other fractions and DMC had no cytotoxicity to the three cancer cell lines (CC_{50} values ranging from 85.43 to >100 µg/mL) or weak cytotoxicity (CC_{50} values ranging from 51.06 to 77.07 µg/mL). However, the four tested cell lines were less sensitive to these fractions and DMC when compared with the positive control camptothecin, with a CC_{50} of 0.005–1.57 µg/mL ($p < 0.001$).

Table 5. Cytotoxicity of *C. operculatus* fractions and DMC against four human cell lines.

Samples	CC_{50} (µg/mL)			
	MCF-7	Jurkat	HeLa	Fibroblast
BHF	30.79 ^{d} ± 0.83	18.51 ^{d} ± 1.37	31.70 ^{c} ± 5.74	>100
BEF	57.74 ^{c} ± 1.13	51.06 ^{c} ± 1.42	92.53 ^{a} ± 4.90	>100
BAF	>100	>100	>100	>100
LHF	89.00 ^{a} ± 1.33	56.78 ^{c} ± 2.62	86.45 ^{ab} ± 4.06	>100
LEF	85.43 ^{a} ± 2.76	66.73 ^{b} ± 1.02	91.72 ^{a} ± 1.43	>100

Table 5. *Cont.*

Samples	CC$_{50}$ (μg/mL)			
	MCF-7	Jurkat	HeLa	Fibroblast
LAF	>100	97.25 ^{a} ± 1.28	77.07 ^{b} ± 5.06	>100
DMC	71.41 ^{b} ± 2.14	73.82 ^{b} ± 7.23	>100	>100
Camptothecin	0.007 ^{e} ± 0.002	0.005 ^{e} ± 0.001	0.89 ^{d} ± 0.088	1.57 ± 0.84

BHF: flower bud hexane fraction, BEF: flower bud ethyl acetate fraction, and BAF: flower bud aqueous fraction. LHF: leaf hexane fraction, LEF: leaf ethyl acetate fraction, and LAF: leaf aqueous fraction. Data are shown as mean ± SD (n = 3). Means within a column with different letters indicate a significant difference at $p < 0.05$.

4. Discussion

Phytochemicals (e.g., polyphenols, flavonoids, alkaloids, tannins, and saponins) are known to play an important role in overall health and disease prevention. In the *C. operculatus* flower bud and leaf fractions, the contents of phytochemicals significantly varied depending on the solvents used. Similarly with other previous studies, a high polyphenol content was mostly found in the aqueous extracts of both *C. operculatus* flower buds and leaves [22,23,37–39]. Minh et al. (2023) [40] recently showed that the flavonoid content was only present at a low level in the aqueous extracts of *C. operculatus* fresh leaves (2.73 mg QE/g) when compared with that in the hexane and ethyl acetate extracts. The flavonoid contents were present at low levels in the aqueous extracts of *C. operculatus* fresh leaves (2.73 mg QE/g) [40] and flower buds (81.1 mg QE/g) [4] when compared with that in the hexane fraction.

In our study, flavonoids have been found to be remarkably rich in the hexane and ethyl acetate fractions. Chalcones belonging to flavonoids family, especially DMC, have been presented as bioactive agents isolated from the non- or low-polar solvent extracts of the flower buds [1,2,9,41,42]. The contents of saponins and tannins were also high in these hexane and ethyl acetate fractions, particularly in the *C. operculatus* flower buds. The tannin content in the hexane fraction was reported to be higher than that in the aqueous extract of *C. nervosum* pulp [43].

It is well known that higher polyphenol and flavonoid contents could significantly increase the antioxidant capacity of plant extracts [44]. In the present study, the hexane and ethyl acetate fractions exhibited weaker antioxidant capacities than the aqueous counterparts. This may be because the content of flavonoids was present at lower levels than those of other phenolic compounds in the *C. operculatus* flower buds and leaves. The aqueous fractions in our study exhibited a similar antioxidant capacity to those of aqueous extracts from *C. operculatus* flower buds and leaves reported by Mai et al. (2009) [37]. Compared to the other studies, most of the *C. operculatus* fractions had a higher antioxidant capacity than those of *Psidium guajava* L. [45], *Halimium halimifolium* [21], and *Syzygium aromaticum* flower buds [22].

The studies by Mai et al. (2009) [37] and Minh et al. (2023) [40] have shown that there were very little to no alkaloids in *C. operculatus.* Our study identified alkaloids in all of the tested fractions at significantly lower concentrations than other phytochemicals. Although the antioxidant activity of alkaloids was also demonstrated [6], the presence of alkaloids at low levels in the fractions of the *C. operculatus* flower buds and leaves could make their contribution secondary when compared with polyphenols and flavonoids.

Among the fractions of *C. operculatus* flower buds and leaves, the hexane fractions and the isolated compound DMC presented the strongest growth inhibitory activity against *H. pylori.* This indicated that flavonoids and other phenolic compounds could have a high synergistic activity against the growth of *H. pylori.* Especially, the single compound DMC, isolated from the flower buds of *C. operculatus*, was found to have two-and-a-half to five times more potent inhibitory activity than these hexane fractions. Similarly, terpenes (e.g., diterpenes and sesquiterpenes) and phenolic compounds from the leaf extract of *Caseria sylvestris* have been reported to provide a high synergism against *H. pylori* [46]. Several studies have shown that extracts from the flower buds and leaves possessed antibacterial activity against many types of Gram-negative and Gram-positive bacteria such as *Xanthomonas* spp. [47], *Staphylococcus aureus, S. epidermidis, S. haemolyticus, Bacillus subtilis, Streptococcus mutans* [6], *H. pylori* [11], and *S. pyogenes* [48]. In addition, DMC has also been known as a bioactive agent against cellular oxidative stress [2] and cytotoxic effects [21,49]. Crude ethanolic extracts from the leaves [11] and crude hexane extracts from the flower buds of *C. operculatus* [4] were previously reported to have high antibacterial activity against *H. pylori*. However, to date, there are no data reported on the antibacterial activity of DMC isolated from *C. operculatus* flower buds against this pathogenic bacterium.

The antibacterial activity of *C. operculatus* fractions against *H. pylori* was found to be higher than those of other reported medicinal plants such as *Cichorium intybus* (MICs of 1.25–10 mg/mL), *Cinnamomum zeylanicum* (MICs of 1.25 to 5 mg/mL), *Foeniculum vulgare* (MICs of >10 mg/mL) [50], and *H. rosa sinensis* (MICs of 0.2–0.25 mg/mL) [28]. It was previously reported that 2′,4′- dihydroxychalcone isolated from leaves of *Muntingia calabura* provided antibacterial activity against methicillin-susceptible and resistant *Staphylococcus aureus* (MICs of 50 and 100 mg/mL, respectively) [51]. Recently, the flavonoids naringenin, myricetin, and luteolin isolated from the red flowers of *H. rosa sinensis* were reported to display effective anti-*H. pylori* properties (MICs of 100–150 μg/mL) [32]. In our study, DMC (2′,4′-Dihydroxy-6′-methoxy-3′,5′-dimethylchalcone) was found to possess potent growth inhibitory activity against *H. pylori* (MICs of 25–50 μg/mL).

H. pylori can produce urease for successful survival in acidic conditions and colonization in the gastric mucosa of the human stomach. Therefore, the inhibition of the bacterial enzyme could prevent bacterial growth and colonization. The crude hexane extract from flower buds of *C. operculatus* was found to have a strong inhibitory effect on *H. pylori* urease activity [9] The *C. operculatus* fractions and DMC in our present study have also been found to possess potent inhibitory effects on the urease of *H. pylori*, and is much stronger than those of *Fagonia arabica* L. and *Casuarina equisetifolia* L. reported by Amin et al. (2013) [52]. The flavonoid quercetin, present in the acetone extract of *Heterotheca inuloides* Cass. (Asteraceae), has been known to display a high in vitro inhibition against the enzyme, with an IC_{50} = 132.4 μg/mL [53]. Several other flavonoids (such as naringenin, myricetin, and luteolin) and phenolic acid (protocatechuic acid) from the red flowers of *Hibiscus rosa sinensis* were also reported to produce the potent inhibition of *H. pylori* urease activity [32].

In addition, inhibition of α-amylase and α-glucosidase activities has been known to be one of the treatments for diabetes, since it helps to control glucose levels in the blood. Previously reported by Zhang and Lu (2012) [8], the aqueous extract of *C. operculatus* flower buds and DMC inhibited α-amylase, with IC_{50} values of 73.10 and 20.67 μg/mL, respectively. Recently, Chukiatsiri et al. (2023) [54] indicated that the hexane extract of *C. nervosum* had no inhibitory activity against both α-amylase and α-glucosidase, but the aqueous extract of this plant revealed a depressing effect on both of the enzymes, with IC_{50} values of 0.61 and 0.44 mg/mL, respectively. In the present study, we found that the *C. operculatus* fractions had a strong inhibitory effect on α-glucosidase, while DMC displayed inhibitory activity against α-amylase.

Moreover, adverse effects on the bacterial biofilm formation, cell morphology, and membrane permeability have also been well described as a mode of action of plant secondary metabolites on bacterial survival [55,56]. In the current study, we found that all *C. operculatus* fractions and DMC inhibited urease, with IC_{50} values significantly smaller than their MIC values. At sub-MICs, they also exhibited pronounced antibiofilm activity, as the biofilm growth can act as reservoirs for the spread of the pathogenic bacterium, persistent infection, and resistance to adverse factors [57]. The EtOAc fraction of *H. rosa-sinensis* red flowers at MIC/2 (0.125 mg/mL) has been reported to inhibit the biofilm formation of *H. pylori* by 79.3% and cause the considerable transformation of the spiral forms to the coccoid forms (91 vs. 18% at 1.5 and 0.75 mg/mL after 48 h of treatment) [28]. The flavanone naringenin was found to display the most antibiofilm activity (85.9 versus 52.7% of inhibition at MIC/2 and MIC/4 = 25 μg/mL) [32]. The flavanone also induced the morphological conversion of *H. pylori* to the coccoid forms (95 vs. 16.5% at 1000 and 500 μg/mL), followed by the flavone luteolin (87.5 vs. 14.7%), which caused a greater change to the coccoid forms than that of the flavonol myricetin (79 vs. 15%) [32]. Previously, the methanolic extract of *C. operculatus* leaves were known to inhibit the ability of the acid production and biofilm formation of *Streptococcus mutans*, leading to its anticaries activity [6]. This current study is first report of *anti-H. pylori* and antibiofilm activities and the morphological conversion of the bacterial cells caused by the *C. operculatus* flower bud and leaf fractions and DMC.

Furthermore, naturally occurring chalcones found in many medicinal and edible plants have been known to be precursors of plant flavonoids [58]. Flavonoids possessing antibacterial properties could penetrate the lipid bilayer membrane, causing an increase in membrane permeability and alleviating the bacterial pathogenicity [59]. In our study, the *C. operculatus* flower bud and leaf fractions, especially the isolated DMC, were proven to effectively increase the membrane permeability of *H. pylori*. The results from the SEM images indicated that the hexane fraction of the *C. operculatus* flower buds and DMC produced extensive morphological damage, causing an increase in the membrane permeability, and exerted bactericidal effects. The *H. pylori* cells treated with hesperetin, naringenin, and 7-O-butylnaringenin were shown to be damaged and resulted in morphological alterations or irregular shapes and rough surfaces [60,61]. Hesperidin was reported to interact with bacterial cells and induce membrane disruption, leading to the leakage of cytoplasmic components prior to cell death [62]. Ergüden and Ünver (2021) [63] proposed that phenolic chalcones caused ion leakage from the cytoplasm of Gram-positive bacteria before membrane deformation.

The cytotoxicity experiments showed that the *C. operculatus* fractions and DMC were not toxic to the tested cell lines, in which the flower bud hexane fraction (DMC-rich fraction) was more toxic to the three cancer cell lines (MCF-7, Jurkat, and HeLa) than to the fibroblast cells. The DMC-rich extract obtained from *S. nervosum* fruits was also reported to have stronger anticancer activities against HepG2 (human liver cancer cells) and A549 (human lung cancer cells) than the isolated DMC [21]. These showed that the fractions and DMC have selective inhibitory effects on *H. pylori* urease and induce morphological conversions and membrane disruption, leading to the cell death.

5. Conclusions

The fractions from the flower buds and leaves of *C. operculatus* exerted various antioxidant activities. The results demonstrated that the aqueous fractions of the flower buds and leaves, which contain high phenolic contents, exhibited significantly antioxidant effects. The flower bud and leaf hexane fractions, with rich contents of flavonoids and saponins, possessed considerable antibacterial activity towards *H. pylori*. The pronounced anti-*H. pylori* activity of the hexane fractions and DMC (2′,4′-dihydroxy-6′-methoxy-3′,5′-dimethylchalcone) may result from its potential role in urease inhibitory and antibiofilm properties. Additionally, the growth-inhibiting and bactericidal effects of the flower bud hexane fractions and DMC have been attributed to causing morphological changes, increasing the permeability, and damaging the cell membrane of *H. pylori*. Moreover, the activities of *C. operculatus* fractions to inhibit α-glucosidase and DMC against α-amylase, and their safety for human cells, indicate the need to evaluate their biological effectiveness in vivo. Flower bud- and leaf-derived materials rich in chalcones and flavonoids, particularly hexane fractions, promise to be a potential source for further pharmaceutical studies.

Supplementary Materials: The following supporting information can be downloaded at: https://www.mdpi.com/article/10.3390/biotech13040042/s1, Figure S1: Scheme of flower bud and leaf solvent fractions partitioned by liquid–liquid fractionation from the flower bud and leaf ethanolic extracts of *Cleistocalyx operculatus*, and the isolation of principle 1 derived from the flower bud hexane fraction (BHF); Figure S2: The HPLC of DMC isolated from the flower bud hexane fraction (BHF); Figure S3: FT-IR data of DMC isolated from the flower bud hexane fraction (BHF); Figure S4: ^{1}H-NMR data of DMC isolated from the flower bud hexane fraction (BHF); Figure S5: ^{13}C-NMR data of DMC isolated from the flower bud hexane fraction (BHF); Figure S6: ^{1}H–^{13}C HMBC-correlations of DMC isolated from the flower bud hexane fraction (BHF); Table S1: ^{1}H NMR, ^{13}C-NMR, and HMBC data of DMC isolated from the flower bud hexane fraction (BHF).

Author Contributions: Conceptualization, D.T.T., L.T.M.N. and T.T.H.; methodology, D.T.T., M.T.T., N.T.M.T., P.N.P.T., P.T.H.T., P.T.G.T., L.T.M.N. and T.T.H.; software, D.T.T.; validation, D.T.T., L.T.M.N. and T.T.H.; formal analysis, D.T.T., M.T.T. and P.T.G.T.; investigation, D.T.T., M.T.T., N.T.M.T., P.N.P.T., P.T.H.T., P.T.G.T., L.T.M.N. and T.T.H.; resources, D.T.T., L.T.M.N. and T.T.H.; data curation, D.T.T., M.T.T. and N.T.M.T.; writing—original draft preparation, D.T.T. and T.T.H.; writing—review and editing, D.T.T. and T.T.H.; supervision, L.T.M.N. and T.T.H.; project administration, L.T.M.N.; funding acquisition, L.T.M.N. All authors have read and agreed to the published version of the manuscript.

Funding: This research was funded by the Vietnam National Foundation for Science and Technology Development (NAFOSTED) under grant number 106-YS.06-2015.17.

Institutional Review Board Statement: The study was approved by the Ethical Committee of the Vietnam National University, Ho Chi Minh City (No. 702/DHQG-KHCN), Vietnam, on 24 April 2017.

Informed Consent Statement: Not applicable.

Data Availability Statement: The data presented in this study are available in this article and Supplementary Materials.

Conflicts of Interest: The authors declare no conflicts of interest.

References

1. Ye, C.-L.; Lu, Y.-H.; Li, X.-D.; Wei, D.-Z.; Whitehead, C. HPLC analysis of a bioactive chalcone and triterpene in the buds of *Cleistocalyx operculatus*. *S. Afr. J. Bot.* **2005**, *71*, 312–315. [CrossRef]
2. Ye, C.L.; Liu, X.G.; Huang, Q. Antioxidant activity and protection of human umbilical vein endothelial cells from hydrogen peroxide-induced injury by DMC, a chalcone from buds of *Cleistocalyx operculatus*. *S. Afr. J. Bot.* **2013**, *86*, 36–40. [CrossRef]
3. Pham, G.N.; Nguyen, T.T.T.; Nguyen-Ngoc, H. Ethnopharmacology, Phytochemistry, and Pharmacology of Syzygium nervosum. *Evid. Based Complement. Altern. Med.* **2020**, *2020*, 8263670. [CrossRef] [PubMed]
4. Thanh, D.T.; Oanh, V.K.; Nguyen, H.C.; Ngan, L.T.M.; Hieu, T.T. Phytochemical composition, antioxidant, antibacterial, and enzyme inhibitory activities of organic extracts from flower buds of *Cleistocalyx operculatus* (Roxb.) Merr. et Perry. *Bio Technol.* **2024**, *105*, 137–147. [CrossRef]
5. Dung, N.T.; Kim, J.M.; Kang, S.C. Chemical composition, antimicrobial and antioxidant activities of the essential oil and the ethanol extract of *Cleistocalyx operculatus* (Roxb.) Merr and Perry buds. *Food Chem. Toxicol.* **2018**, *46*, 3632–3639. [CrossRef] [PubMed]
6. Nguyen, P.T.M.; Schultze, N.; Boger, C.; Alresley, Z.; Bolhuis, A.; Lindequist, U. Anticaries and antimicrobial activities of methanolic extract from leaves of *Cleistocalyx operculatus* L. *Asian Pac. J. Trop. Biomed.* **2017**, *7*, 43–48. [CrossRef]
7. Woo, A.Y.H.; Waye, M.M.Y.; Kwan, H.S.; Chan, M.C.Y.; Chau, C.F.; Cheng, C.H.K. Inhibition of ATPases by *Cleistocalyx operculatus*. A possible mechanism for the cardiotonic actions of the herb. *Vasc. Pharmacol.* **2002**, *38*, 163–168. [CrossRef]
8. Zhang, L.; Lu, Y. Inhibitory activities of extracts from *Cleistocalyx operculatus* flower buds on pancreatic lipase and α-amylase. *Eur. Food Res. Technol.* **2012**, *235*, 1133–1139. [CrossRef]
9. Ye, C.L.; Lu, Y.H.; Wei, D.Z. Flavonoids from *Cleistocalyx operculatus*. *Phytochemistry* **2004**, *65*, 445–447. [CrossRef]
10. Gafner, S.; Wolfender, J.L.; Mavi, S.; Hostettmann, K. Antifungal and antibacterial chalcones from Myrica serrata. *Planta Medica* **1996**, *62*, 67–69. [CrossRef]
11. Ngan, L.T.M.; Dung, P.P.; Nhi, N.V.T.Y.; Hoang, N.V.M.; Hieu, T.T. Antibacterial activity of ethanolic extracts of some Vietnamese medicinal plants against *Helicobacter pylori*. *AIP Conf. Proc.* **2017**, *1878*, 020030. [CrossRef]
12. Elbehiry, A.; Marzouk, E.; Aldubaib, M.; Abalkhail, A.; Anagreyyah, S.; Anajirih, N.; Almuzaini, A.M.; Rawway, M.; Alfadhel, A.; Draz, A.; et al. *Helicobacter pylori* Infection: Current Status and Future Prospects on Diagnostic, Therapeutic and Control Challenges. *Antibiotics* **2023**, *12*, 191. [CrossRef] [PubMed]
13. Wroblewski, L.E.; Peek, R.M.; Wilson, K.T. *Helicobacter pylori* and gastric cancer: Factors that modulate disease risk. *Clin. Microbiol. Rev.* **2010**, *23*, 713–739. [CrossRef] [PubMed]
14. Tsugane, S.; Sasazuki, S. Diet and the risk of gastric cancer: Review of epidemiological evidence. *Gastric Cancer* **2007**, *10*, 75–83. [CrossRef]
15. Eusebi, L.H.; Zagari, R.M.; Bazzoli, F. Epidemiology of *Helicobacter pylori* Infection. *Helicobacter* **2014**, *19*, 1–5. [CrossRef]
16. Kusters, J.G.; Van Vliet, A.H.M.; Kuipers, E.J. Pathogenesis of *Helicobacter pylori* infection. *Clin. Microbiol. Rev.* **2006**, *19*, 449–490. [CrossRef]
17. Malfertheiner, P.; Camargo, M.C.; El-Omar, E.; Liou, J.M.; Peek, R.; Schulz, C.; Smith, S.I.; Suerbaum, S. *Helicobacter pylori* infection. *Nat. Rev. Dis. Primers* **2023**, *9*, 19. [CrossRef]
18. Yonezawa, H.; Osaki, T.; Kamiya, S. Biofilm formation by *Helicobacter pylori* and its involvement for antibiotic resistance. *BioMed Res. Int.* **2015**, *2015*, 914791. [CrossRef] [PubMed]

19. Gaddy, J.A.; Radin, J.N.; Cullen, T.W.; Chazin, W.J.; Skaar, E.P.; Trent, M.S.; Algood, H.M.S. *Helicobacter pylori* resists the antimicrobial activity of calprotectin via lipid a modification and associated biofilm formation. *mBio* **2015**, *6*, e01349-15. [CrossRef]
20. Nguyen, M.N.T.; Huynh, T.D.H. Selective cytotoxicity of a Vietnamese traditional formula, Nam Dia long, against MCF-7 cells by synergistic effects. *BMC Complement. Altern. Med.* **2016**, *16*, 220. [CrossRef]
21. Choommongkol, V.; Punturee, K.; Klumphu, P.; Rattanaburi, P.; Meepowpan, P.; Suttiarporn, P. Microwave-Assisted Extraction of Anticancer Flavonoid, 2′,4′-Dihydroxy-6′-methoxy-3′,5′-dimethyl Chalcone (DMC), Rich Extract from Syzygium nervosum Fruits. *Molecules* **2022**, *27*, 1397. [CrossRef] [PubMed]
22. Temesgen, S.; Sasikumar, J.M.; Egigu, M.C. Effect of Extraction Solvents on Total Polyphenolic Content and Antioxidant Capacity of *Syzygium Aromaticum* L. Flower Bud from Ethiopia. *BioMed Res. Int.* **2022**, *2022*, 4568944. [CrossRef] [PubMed]
23. Rebaya, A.; Belghith, S.I.; Baghdikian, B.; Leddet, V.M.; Mabrouki, F.; Olivier, E.; Cherif, J.K.; Ayadi, M.T. Total Phenolic, Total Flavonoid, Tannin Content, and Antioxidant Capacity of *Halimium halimifolium* (Cistaceae). *J. Appl. Pharm. Sci.* **2015**, *5*, 52–57. [CrossRef]
24. Ncube, B.; Nair, J.J.; Rárová, L.; Strnad, M.; Finnie, J.F.; Van Staden, J. Seasonal pharmacological properties and alkaloid content in Cyrtanthus contractus N.E. *Br. S. Afr. J. Bot.* **2015**, *97*, 69–76. [CrossRef]
25. Benyong, H.; Ying, C.; Ying, R.; Chaoyin, C. Content determination of total saponins from Opuntia. *Bio Technol. Indian J.* **2014**, *10*, 2014. Available online: https://www.tsijournals.com/articles/content-determination-of-total-saponins-from-opuntia.pdf (accessed on 15 January 2024).
26. Elouafy, Y.; El Yadini, A.; Mortada, S.; Hnini, M.; Harhar, H.; Khalid, A.; Abdalla, A.; Bouyahya, A.; Goh, K.; Ming, L.; et al. Antioxidant, antimicrobial, and α-glucosidase inhibitory activities of saponin extracts from walnut (*Juglans regia* L.) leaves. Asian Pac. *J. Trop. Biomed.* **2023**, *13*, 60–69. [CrossRef]
27. Olszowy-Tomczyk, M.; Typek, R. Transformation of phenolic acids during radical neutralization. *J. Food Sci. Technol.* **2024**, *61*, 790–797. [CrossRef] [PubMed]
28. Ngan, L.T.M.; Tan, M.T.; Hoang, N.V.M.; Thanh, D.T.; Linh, N.T.T.; Hoa, T.T.H.; Nuong, N.T.M.; Hieu, T.T. Antibacterial activity of Hibiscus rosa-sinensis l. red flower against antibiotic-resistant strains of *Helicobacter pylori* and identification of the flower constituents. *Braz. J. Med Biol. Res.* **2021**, *54*, e10889. [CrossRef] [PubMed]
29. Ngan, L.T.M.; Moon, J.K.; Kim, J.H.; Shibamoto, T.; Ahn, Y.J. Growth-inhibiting effects of *Paeonia lactiflora* root steam distillate constituents and structurally related compounds on human intestinal bacteria. *World J. Microbiol. Biotechnol.* **2012**, *28*, 1575–1583. [CrossRef]
30. Shai, L.J.; Magano, S.R.; Lebelo, S.L.; Mogale, A.M. Inhibitory effects of five medicinal plants on rat alpha-glucosidase: Comparison with their effects on yeast alpha-glucosidase. *J. Med. Plant. Res.* **2011**, *5*, 2863–2867.
31. Ogunyemi, O.M.; Gyebi, G.A.; Saheed, A.; Paul, J.; Nwaneri-Chidozie, V.; Olorundare, O.; Adebayo, J.; Koketsu, M.; Aljarba, N.; Alkahtani, S.; et al. Inhibition mechanism of alpha-amylase, a diabetes target, by a steroidal pregnane and pregnane glycosides derived from *Gongronema latifolium* Benth. *Front. Mol. Biosci.* **2022**, *9*, 866719. [CrossRef] [PubMed]
32. Hieu, T.T.; Truong, T.H.H.; Nguyen, T.T.L.; Nguyen, V.M.H.; Nguyen, T.M.N.; Luong, T.M.N. Growth-inhibiting, bactericidal, antibiofilm, and urease inhibitory activities of *Hibiscus rosa sinensis* L. flower constituents toward antibiotic sensitive- and resistant-strains of *Helicobacter pylori*. *ACS Omega* **2020**, *5*, 20080–20089. [CrossRef]
33. Guzman, J.; Téné, N.; Touchard, A.; Castillo, D.; Belkhelfa, H.; Haddioui-Hbabi, L.; Treilhou, M.; Sauvain, M. Anti-*Helicobacter pylori* properties of the ant-venom peptide bicarinalin. *Toxins* **2018**, *10*, 21. [CrossRef]
34. Devi, K.P.; Nisha, S.A.; Sakthivel, R.; Pandian, S.K. Eugenol (an essential oil of clove) acts as an antibacterial agent against Salmonella typhi by disrupting the cellular membrane. *J. Ethnopharmacol.* **2010**, *130*, 107–115. [CrossRef] [PubMed]
35. Schwarz, S.; Böttner, A.; Goosens, L.; Hafez, H.M.; Hartmann, K.; Kaske, M.; Kehrenberg, C.; Kietzmann, M.; Klarmann, D.; Klein, G.; et al. A proposal of clinical breakpoints for amoxicillin applicable to porcine respiratory tract pathogens. *Vet. Microbiol.* **2008**, *126*, 178–188. [CrossRef] [PubMed]
36. Ngan, L.T.M.; Vi, N.T.; Tham, D.T.M.; Loan, L.T.T.; Ho, P.T.; Hieu, T.T. Antioxidant and anti-*Helicobacter pylori* activities of Hericium erinaceus mycelium and culture filtrate. *Biomed. Res. Ther.* **2021**, *8*, 4266–4275. [CrossRef]
37. Mai, T.T.; Fumie, N.; Chuyen, N.V. Antioxidant activities and hypolipidemic effects of an aqueous extract from flower buds of *Cleistocalyx operculatus* (Roxb.) Merr. and Perry. *J. Food Biochem.* **2009**, *33*, 790–807. [CrossRef]
38. Seo, J.; Lee, S.; Elam, M.L.; Johnson, S.A.; Kang, J.; Arjmandi, B.H. Study to find the best extraction solvent for use with guava leaves (*Psidium guajava* L.) for high antioxidant efficacy. *Food Sci. Nutr.* **2014**, *2*, 174–180. [CrossRef]
39. Wijekoon, M.M.J.O.; Bhat, R.; Karim, A.A. Effect of extraction solvents on the phenolic compounds and antioxidant activities of bunga kantan (Etlingera elatior Jack.) inflorescence. *J. Food Compost. Anal.* **2011**, *24*, 615–619. [CrossRef]
40. Minh, T.T.L.; Kieu, L.T.B.; Mai, S.T.T.; Ngoc, D.L.B.; Thuy, L.T.B.; Quyen, N.T.; Anh, T.T.; Huy, L.V.; Phong, N.V.; Duyen, C.T.M.; et al. Addition of Mentha arvensis in infusions of *Cleistocalyx operculatus* improves the Hedonic Score and retains the high antioxidant and anti-lipid-peroxidation effects. *Appl. Sci.* **2023**, *13*, 2873. [CrossRef]
41. Dung, N.X.; Van, L.H.; Khoi, T.T.; Leclercq, P.A. GC and GC/MS analysis of the leaf oil of *Cleistocalyx operculatus* Roxb. Merr. et Perry (Syn. Eugenia operculata Roxb.; Syzygicum mervosum DC.). *J. Essent. Oil Res.* **1994**, *6*, 661–662. [CrossRef]
42. Min, B.S.; Thu, C.V.; Dat, N.T.; Nguyen, C.; Dang, H.; Jang, H.S.; Hung, T.M. Antioxidative Flavonoids from *Cleistocalyx operculatus* Buds. *Chem. Pharm. Bull.* **2008**, *56*, 1725–1728. [CrossRef] [PubMed]

43. Charoensin, S.; Taya, S.; Wongpornchai, S.; Wongpoomchai, R. Assessment of genotoxicity and antigenotoxicity of an aqueous extract of *Cleistocalyx nervosum* var. paniala in in vitro and in vivo models. *Interdiscip. Toxicol.* **2012**, *5*, 201–206. [CrossRef]
44. Saeed, N.; Khan, M.R.; Shabbir, M. Antioxidant activity, total phenolic and total flavonoid contents of whole plant extracts *Torilis leptophylla* L. *BMC Complement. Altern. Med.* **2012**, *12*, 221. [CrossRef]
45. Thaipong, K.; Boonprakob, U.; Crosby, K.; Cisneros-Zevallos, L.; Hawkins Byrne, D. Comparison of ABTS, DPPH, FRAP, and ORAC assays for estimating antioxidant activity from guava fruit extracts. *J. Food Compos. Anal.* **2006**, *19*, 669–675. [CrossRef]
46. Spósito, L.; Oda, F.B.; Vieira, J.H.; Carvalho, F.A.; dos Santos Ramos, M.A.; de Castro, R.C.; Crevelin, E.J.; Crotti, A.E.M.; Santos, A.G.; da Silva, P.B.; et al. In vitro and in vivo anti-*Helicobacter pylori* activity of *Casearia sylvestris* leaf derivatives. *J. Ethnopharmacol.* **2019**, *233*, 1–12. [CrossRef]
47. Bajpai, V.K.; Dung, N.T.; Suh, H.J.; Kang, S.C. Antibacterial activity of essential oil and extracts of *Cleistocalyx operculatus* buds against the bacteria of *Xanthomonas* spp. *J. Am. Oil Chem. Soc.* **2010**, *87*, 1341–1349. [CrossRef]
48. Thuy, B.T.P.; Hieu, L.T.; My, T.T.A.; Hai, N.T.T.; Loan, H.T.P.; Thuy, N.T.T.; Triet, N.T.; Van Anh, T.T.; Dieu, N.T.X.; Quy, P.T.; et al. Screening for Streptococcus pyogenes antibacterial and Candida albicans antifungal bioactivities of organic compounds in natural essential oils of *Piper betle* L., *Cleistocalyx operculatus* L. and *Ageratum conyzoides* L. *Chem. Pap.* **2021**, *75*, 1507–1519. [CrossRef]
49. Utama, K.; Khamto, N.; Meepowpan, P.; Aobchey, P.; Kantapan, J.; Sringarm, K.; Roytrakul, S.; Sangthong, P. Effects of 2′,4′-Dihydroxy-6′-methoxy-3′,5′-dimethylchalcone from Syzygium nervosum Seeds on Antiproliferative, DNA Damage, Cell Cycle Arrest, and Apoptosis in Human Cervical Cancer Cell Lines. *Molecules* **2022**, *27*, 1154. [CrossRef]
50. Nostro, A.; Cellini, L.; Di Bartolomeo, S.; Di Campli, E.; Grande, R.; Cannatelli, M.A.; Marzio, L.; Alonzo, V. Antibacterial effect of plant extracts against *Helicobacter pylori*. *Phytother. Res.* **2005**, *19*, 198–202. [CrossRef]
51. Sufian, A.S.; Ramasamy, K.; Ahmat, N.; Zakaria, Z.A.; Yusof, M.I.M. Isolation and identification of antibacterial and cytotoxic compounds from the leaves of *Muntingia calabura* L. *J. Ethnopharmacol.* **2013**, *146*, 198–204. [CrossRef] [PubMed]
52. Amin, M.; Anwar, F.; Naz, F.; Mehmood, T.; Saari, N. Anti-*Helicobacter pylori* and urease inhibition activities of some traditional medicinal plants. *Molecules* **2013**, *18*, 2135–2149. [CrossRef]
53. Egas, V.; Salazar-Cervantes, G.; Romero, I.; Méndez-Cuesta, C.A.; Rodríguez-Chávez, J.L.; Delgado, G. Anti-*Helicobacter pylori* metabolites from Heterotheca inuloides (*Mexican arnica*). *Fitoterapia* **2018**, *127*, 314–321. [CrossRef]
54. Chukiatsiri, S.; Wongsrangsap, N.; Ratanabunyong, S.; Choowongkomon, K. In vitro Evaluation of Antidiabetic Potential of *Cleistocalyx nervosum* var. paniala Fruit Extract. *Plants* **2023**, *12*, 112. [CrossRef] [PubMed]
55. Wink, M. Importance of plant secondary metabolites for protection against insects and microbial infections. *Adv. Phytomed.* **2006**, *3*, 251–268. [CrossRef]
56. Tran, T.H.; Luong, T.M.N.; Bui, V.L.; Tran, T.H. Effects of plant essential oils and their constituents on *Helicobacter pylori*: A Review. *Plant Sci. Today* **2023**, *10*, 334–344. [CrossRef]
57. Singh, A.; Amod, A.; Pandey, P.; Bose, P.; Pingali, S.; Shivalkar, S.; Varadwaj, P.; Sahoo, A.; Samanta, S. Bacterial biofilm infections, their resistance to antibiotics therapy and current treatment strategies. *Biomed. Mater.* **2022**, *17*, 022003. [CrossRef]
58. Rudrapal, M.; Khan, J.; Dukhyil, A.A.B.; Alarousy, R.M.I.I.; Attah, E.I.; Sharma, T.; Khairnar, S.J.; Bendale, A.R. Chalcone Scaffolds, Bioprecursors of Flavonoids: Chemistry, Bioactivities, and Pharmacokinetics. *Molecules* **2021**, *26*, 7177. [CrossRef]
59. Xie, Y.; Yang, W.; Tang, F.; Chen, X.; Ren, L. Antibacterial Activities of Flavonoids: Structure-Activity Relationship and Mechanism. *Curr. Med. Chem.* **2014**, *22*, 132–149. [CrossRef]
60. Yang, J.S.; Wang, C.M.; Su, C.H.; Ho, H.C.; Chang, C.H.; Chou, C.H.; Hsu, Y.M. Eudesmin attenuates *Helicobacter pylori*-induced epithelial autophagy and apoptosis and leads to eradication of *H. pylori* infection. *Exp. Ther. Med.* **2018**, *15*, 2388–2396. [CrossRef]
61. Moon, S.H.; Lee, J.H.; Kim, K.T.; Park, Y.S.; Nah, S.Y.; Ahn, D.U.; Paik, H.D. Antimicrobial effect of 7-O-butylnaringenin, a novel flavonoid, and various natural flavonoids against *Helicobacter pylori* strains. *Int. J. Environ. Res. Public Health* **2013**, *10*, 5459–5469. [CrossRef] [PubMed]
62. Sharaf, M.; Arif, M.; Hamouda, H.I.; Khan, S.; Abdalla, M.; Shabana, S.; Rozan, H.E.; Khan, T.U.; Chi, Z.; Liu, C. Preparation, urease inhibition mechanisms, and anti-*Helicobacter pylori* activities of hesperetin-7-rhamnoglucoside. *Curr. Res. Microb. Sci.* **2022**, *3*, 100103. [CrossRef] [PubMed]
63. Ergüden, B.; Ünver, Y. Phenolic chalcones lead to ion leakage from Gram-positive bacteria prior to cell death. *Arch. Microbiol.* **2021**, *204*, 3. [CrossRef] [PubMed]

BioTech

Review

Honey Bioactive Molecules: There Is a World Beyond the Sugars

Gregorio Bonsignore, Simona Martinotti † and Elia Ranzato *,†

Dipartimento di Scienze e Innovazione Tecnologica (DiSIT), University of Piemonte Orientale, 15121 Alessandria, Italy; gregorio.bonsignore@uniupo.it (G.B.); simona.martinotti@uniupo.it (S.M.)
* Correspondence: elia.ranzato@uniupo.it; Tel.: +39-0131360260
† These authors contributed equally to this work.

Abstract: Honey's many bioactive compounds have been utilized historically to cure infectious diseases. Beneficial effects are its antiviral, antibacterial, anti-inflammatory, antioxidant, and immune-stimulating qualities. The bee species, geographic location, botanical origin, harvest season, processing, and storage conditions all affect honey's potential for therapeutic use. Honey contains a number of antioxidants and active compounds, such as polyphenols, which have been shown to have disease-preventive properties. Based on their origins, categories, and functions, the main polyphenols found in various honey varieties are examined in this review.

Keywords: antioxidant; anti-inflammatory; immunomodulatory; honey; polyphenols; antibacterial activity; wound repair

Key Contribution: Polyphenols, one of honey's most important ingredients, are among the active chemicals and antioxidants it contains.

1. Introduction

Since ancient times, honey has been used traditionally to treat infectious ailments [1]. Because so many different bioactive chemicals are involved, honey's extraordinary qualities are complex. Honey is steadily gaining credibility as a reliable and efficient therapeutic agent to both the general public and conventional medical professionals [2–4].

Honey's antibacterial, anti-inflammatory, antiviral, antioxidant, and immune-stimulating properties have all been linked to its positive effects (Figure 1). [5]. Honey has been demonstrated to effectively cure a variety of other illnesses, such as arthritis, gastric ulcers, diarrhea, dermatitis, and diabetes, and for wound healing [6].

Honey can also be used to disinfect wounds and cleanse skin [7]. Honey's high sugar content, acidity, hydrogen peroxide content, and other non-peroxide molecules are the majority of the known reasons that give honey these characteristics [8–11].

Researchers have just rediscovered honey's inherent antibacterial qualities. The growing trend of antibiotic resistance and the lack of promising new treatments has stoked interest in honey's potential antibacterial properties.

Honey has been shown in numerous studies to have antibacterial capabilities. Even though honey is produced all over the world, its medicinal qualities can differ and are primarily based on the bee species involved, its geographic region, and its botanical origin [12]. Harvesting season, processing, storage conditions, and environmental considerations are external features that might be involved.

Honey's medicinal potential is quite complicated because of the action of different substances, in addition to the significant differences in these substances' quantities between honeys. The medicinal qualities of honey may be impacted by a number of things, like heat and light exposure [13–15].

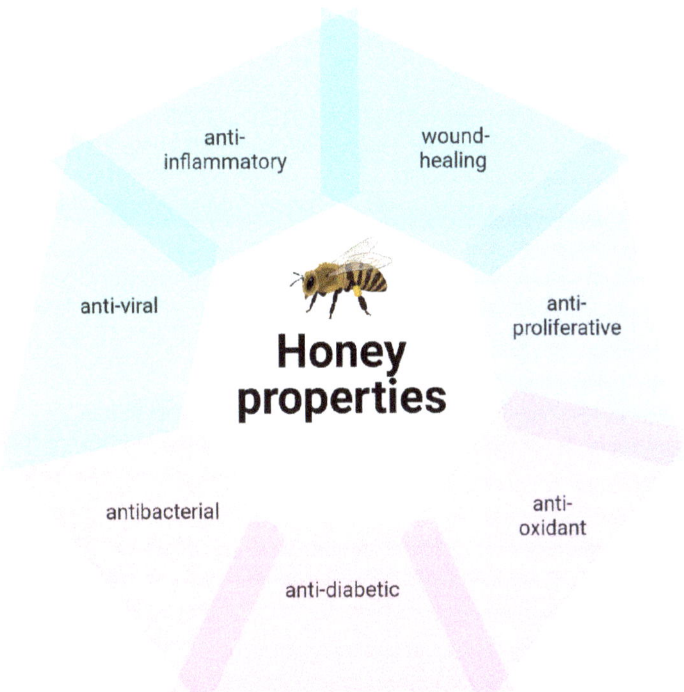

Figure 1. Main recognized properties of honey. For more information, see the text. Created in BioRender. https://BioRender.com/s21r904 (accessed on 11 November 2024).

In fact, numerous active ingredients and antioxidants, including polyphenols, are found in honey. Phytochemicals, a general word for the thousands of plant-based compounds with antioxidant qualities, include polyphenols. The preventive effects of polyphenols against various diseases have been confirmed by several pieces of animal research and numerous in vitro and in vivo experiments. However, it is difficult to pinpoint the precise biological process that underlies each polyphenol and to ascertain the effects of polyphenols on human health [16].

In this review, we explore the most abundant polyphenols found in different kinds of honey based on their source, classification, and specialized roles.

2. Honey and Humankind

Natural honey has the highest energy density of any food. Therefore, it should come as no surprise that honey is a staple food for practically all hunter-gatherers in areas where it is found.

There is also evidence from Stone Age drawings suggesting that honeybee remedies for ailments date back more than 8000 years. Historical documents including old books, tablets, and parchments have documented the use of honey as medicine: Sumerian chalcolithic tablets (6200 BC), the 5000-year-old Veda (a sacred scripture in Hinduism), the Holy Bible, the Koran, and Hippocrates (460–357 BC), as well as Egyptian papyri (1900–1250 BC) [17,18].

2.1. Honey as an Antimicrobial Agent

Honey has demonstrated its efficaciousness as an antibacterial agent against both pathogenic and non-pathogenic microorganisms, including bacteria, yeasts, and fungi, even against those that have evolved a resistance to a variety of antibiotics. Honey has antimicrobial properties that can be either bacteriostatic or bactericidal, depending on the dosage [13,19]. Despite this, several factors, including high osmolarity (low water action), low pH (acidity), hydrogen peroxide (H_2O_2), and non-peroxide components have been linked to its potentials [20–22].

Furthermore, supersaturated sugars, which have a strong affinity for water molecules, are present in bee honey. As a result, bacteria eventually become dehydrated and perish [19].

According to Fahim and coworkers, honey's natural acidity inhibits the growth of microorganisms [23]. Nevertheless, the primary antibacterial activity has been attributed to hydrogen peroxide activity, which is a by-product of the glucose-oxidase enzyme oxidizing glucose, particularly in diluted honey form. Hydrogen peroxide breaks down into very reactive free radicals, which react by killing microorganisms. In general, the honey's ability to retain its properties can be readily interrupted by heat or catalase activity [24].

Honey's antibacterial activity may not be due to peroxides, but rather to non-peroxides molecules, resulting in more consistent and long-lasting benefits.

The oligosaccharides found in honey have additional significant impacts. They have many of the same prebiotic characteristics as fructo-oligosaccharides. Some data have indicated that oligosaccharides are responsible for the increase in population of some advantageous microorganisms responsible for maintaining a healthy gut microbiota in humans, such as *Lactobacilli* and *Bifidobacteria*. It turns out that *Lactobacillus* species protect the body from certain infections like *Salmonellosis*, and *Bifidobacterium* species prevent yeasts or bacterial pathogens from growing too much in the intestinal wall. This may reduce the risk of colon cancer by outcompeting putrefactive microbes that can release agents that cause cancer.

Historical records indicate that honey has long been used as a traditional treatment for microbial diseases [25]. Manuka (*L. scoparium*) honey has been documented to be beneficial against a variety of human infections, including *Enterobacter aerogenes*, *Staphylococcus aureus*, *Salmonella typhimurium*, and *Escherichia coli* [21].

Honey has been shown in several tests to be highly efficient against streptococci, vancomycin-resistant enterococci, and methicillin-resistant *S. aureus* (MRSA) [3]. Nevertheless, due to their greater selectivity against medically significant organisms and their enhanced antibacterial potential for local production (hence accessibility), research has recently discovered that bee honeys may be superior to or similar to manuka honey [20].

2.2. Immunomodulatory and Anti-Inflammatory Properties

Although inflammation is an essential component of the body's normal reaction to infection or damaged tissues, excessive or delayed inflammation can impede recovery or even worsen the situation.

The generation of free radicals within the tissue is the most dangerous consequence of mild inflammation. Certain leucocytes that are stimulated as a major part of the inflammatory process are responsible for initiating these unfastened radicals. Inflammatory processes set off a chain reaction of cellular events that lead to the initiation of growth factors that affect the proliferation of fibroblasts, angiogenesis, and epithelial cells [26].

A number of honey varieties from various nations, including those made by stingless bees, have been claimed to have anti-inflammatory properties [13]. Bee honey's anti-inflammatory properties are a result of its high content of phenolic components. These phenolic and flavonoid chemicals promote the inhibition of the pro-inflammatory effects of cyclooxygenase-1 and cyclooxygenase-2 (COX-1 and COX-2) and/or inducible nitric oxide synthase (iNOS) [27].

Furthermore, the content of prostaglandins, such as prostaglandin E2 (PGE2), thromboxane B2 (in the plasma of healthy individuals), and prostaglandin F2α (PGF2α), decreases when diluted raw bee honey is consumed [15]. There are findings on how various forms of honey can induce the production of interleukin 1β (IL-1β), IL-6, and tumor necrosis factor-alpha (TNF-α) [28,29].

It has also been demonstrated that various honeys, including Gelam honey, lessen inflammatory response mediators like TNF-α and COX-2 by impairing Nuclear Factor kappa B (NF-κB) translocation to the nucleus and blocking the NF-κB pathway. It is commonly recognized that NF-κB activation plays a crucial role in the pathophysiology of inflammation. The sluggish absorption of honey is thought to be the cause of the synthesis of fermentation agents such short-chain fatty acid (SCFA), which has been shown to have immunomodulatory properties. This indicates that some honey-derived fermentable sugars, including nigero-oligosaccharides, have the capacity to trigger an immunological reaction [19,30].

Furthermore, honey contains non-sugar compounds that may be in charge of immunomodulation [31]. Similarly, the topical administration of honey has been discovered to reduce the amount of edema and exudate in wounds, both of which are linked to the activity of the local inflammatory process in wounds [32,33].

2.3. Antioxidant Properties

The antioxidant activity of bee honey in the human body is attributed to its capability to reduce oxidative processes, as measured by its capacity to scavenge free radicals [34]. Honey's anti-inflammatory properties might at least be partially attributed to its antioxidant action, as oxygen free radicals are implicated in a number of inflammatory components [29].

Various flavonoids (such as chrysin, pinocembrin, quercetin, apigenin, and galangin), ascorbic acid, Maillard reaction products, and peptides are found naturally in honey.

Moreover, in honey we can also find tocopherols, catalase, superoxide dismutase, reduced glutathione, and phenolic acids (such as ferulic, ellagic, caffeic, and p-coumaric acids), the majority of which act in concert to provide a synergistic antioxidant action [35, 36].

Honey's antioxidant effect is achieved by inhibiting the production of free radicals, a process that is typically aided by metal ions such as iron, copper, etc. Some typical components of honey, such as flavonoids and other similar polyphenols, may be able to capture these metal ions in complexes, preventing the initial generation of free radicals [29].

3. Honey Composition

About 200 different chemicals may be found in honey, the majority of which are sugars, water, and other components including proteins (enzymes) [37].

The substances extracted from honey harvesting include organic acids, minerals (iron, copper, magnesium, calcium, potassium, phosphorus, zinc, and sodium), vitamins (particularly B6, thiamine, niacin, riboflavin, and pantothenic acid), pigments, phenolic compounds, and a wide range of volatile compounds [38–40].

3.1. Sugars

About 75% of the sugars in honey are monosaccharides, with the remaining 10–15% being disaccharides and trace quantities of other sugars. The attributes of honey, including its energy value, viscosity, hygroscopicity, and granulation, are attributed to its sugar content [41]. The primary factors influencing a honey's sugar content are its geographical and botanical origin, climate, processing, and storage [42,43].

The ratio between the concentrations of fructose and glucose can be used as an indication to help categorize monofloral honeys [44]. Fructose makes up the majority of the carbohydrates in practically all varieties of honey, with the exception of some, such dandelion (*Taraxacum officinale*) and rape (*Brassica napus*), whereas the proportion of glucose may be larger than the fraction of fructose; as a result, these honeys typically crystallize quickly [42].

Honey's sugars and other ingredients can alter when it is being stored. Pentoses and hexoses break down into unfavorable compounds like furans when honey is cooked or kept for an extended period of time. This process involves a gradual enolization and a quick β-elimination of three molecules of water [45]. Furfural, which comes from pentoses, and 5-

hydroxymethylfurfural (5-HMF), which comes from hexoses like glucose and fructose, are the two primary furans that are produced [46]. These are the principal by-products of sugar degradation, and the presence of these compounds in food is typically associated with non-enzymatic browning events, such as caramelization, the Maillard reaction, and sugar breakdown in an acidic media. Actually, these furans have been employed as indicators for the heat treatment of food [46]. In addition to the compounds listed above, heat in the presence of amino acids can also produce other products of sugar degradation, such as 2-acetylfuran [47], isomaltol [48], 3,5-di hydroxy-2-methyl-5,6-diidropiran-4-one, and maltol. These products can alter the color, taste, and odor of honey.

3.2. Proteins

Depending on the species of honeybee, honey shows a different protein composition. One percent (w/w) of honey's ingredients are amino acids, and the amounts of these components vary depending on the honey's source (nectar vs. honeydew) [49].

According to study [50], the protein content of *Apis cerana* and *Apis mellifera* honey ranges from 0.1% to 3.3% and 0.2% to 1.6%, respectively. Pollen is the primary source of protein in honeys; however, proteins and amino acids can also be found in animal and plant sources, such as fluids and nectar secretions from the salivary glands and pharynx of honeybees [37,51].

Proline is the most prevalent amino acid found in pollen and honey [52], accounting for between 50 and 85% of the amino acids in honey [52,53].

Proline has been used as a criterion to assess honey's maturity and, in certain situations, to detect sugar adulteration. For pure honey, the recommended limit value for proline is 180 mg kg^{-1} at the very least [49,54].

Honey contains the enzyme glucose oxidase. After being hydrolyzed to produce gluconic acid, it transforms glucose into D-gluconolactone. In addition to gluconolactone, hydrogen peroxide—which has bactericidal properties, as well as others—is produced by glucose oxidase [55].

Because of the interaction between the free amino groups and the carboxylic group on the reducing end of sugars, storage and processing conditions may result in the creation of unwanted compounds between proteins and amino acids (Maillard reaction).

After six months of storage, the concentration of proline fell, and after twelve months, the concentration of aspartic acid reduced. Between 6 and 24 months, there was no change in the concentration of β-alanine. This finding could indicate that, although β-alanine participates in the Maillard process, it is less reactive than other amino acids [56].

3.3. Organic Acids

Honey has an approximate organic acid content of 0.5% of its fresh weight [57]. Despite being few in number, organic acids are crucial to the physical, chemical, and organoleptic characteristics of honey [58]. Additionally, organic acids can be employed as markers to determine the genuineness and freshness of honey [59]. According to reports, organic acids support honey's antioxidant and antibacterial properties [60].

These organic acids are generated from sugars by enzymes released by bees either directly from nectar or during the process of turning it into honey [61]. These acids have an impact on honey's color, flavor, and chemical characteristics [57].

Honey contains a variety of organic acids depending on the geographical area, including aspartic acid, acetic, formic, butyric, citric, fumaric, galacturonic, gluconic, glutamic, and glutaric acid, lactic, malic, malonic, methylmalonic, 2-oxopentanoic, propionic, pyruvic, quinic, shikimic, succinic, tartaric, oxalic, and other acids [57,61,62].

Gluconic acid is the main acid found in honey. The source of this compound is glucose oxidase, which bees supply to honey throughout the ripening process [63]. Apart from gluconic acid, since honey also contains citric acid, the citric acid content of honey can be utilized as a trustworthy indicator to distinguish floral honey from honeydew [57].

3.4. Vitamins

Tiny levels of vitamins, particularly the B complex, which are derived from suspended pollen grains, can be found in honey. Thiamine (B1), riboflavin (B2), nicotinic acid (B3), pantothenic acid (B5), pyridoxine (B6), biotin (B8 or H), and folic acid (B9) are among the vitamins that can be found in honey.

Vitamin C, which has mostly been researched for its antioxidant qualities, is present in almost all types of honey. Vitamin C is regarded as an unstable indication due to its high susceptibility to chemical and enzymatic oxidation and its rapid changes in reaction to light, oxygen, and heat [64].

3.5. Minerals

Numerous chemical component groups have been identified in many varieties of honey. Some minerals, including potassium, magnesium, and calcium as well as macro- and microelements, silver, zinc, lithium, cobalt, nickel, cadmium, copper, barium, chromium, selenium, sodium, manganese, iodine, zinc, iron, and phosphorus are present in various honeys [38], making up those groups that are significant for the human diet [65].

According to Alqarni et al. [38], the mineral concentration of honey varies from 0.04% in light honeys to 0.2% in dark honeys.

The amount of trace elements in honey changes depending on the type of soil the plant and its nectar were found in, because honeybees' collection of plants' chemical components is reflected in the honey [37,65].

When Nalda and colleagues [66] identified minerals in several Spanish honeys, they discovered that honey made from heather and ling flowers had a greater manganese content. Ajtony and colleagues [66] assessed the mineral composition of honey samples from Hungary and discovered that linden honey had the lowest concentrations of lead, copper, chromium, and cadmium, while these elements were present in higher concentrations in other varieties of honey.

The most prevalent element in honeys is normally potassium, which makes up almost one-third of all the minerals in honey [38].

Mineral elements are not degraded by exposure to heat, light, oxidizing agents, high pH, or other circumstances that damage organic nutrients.

3.6. Polyphenols

Polyphenols represent a diverse group of chemical substances that can be further classified into non-flavonoids (phenolic acids) and flavonoids (flavonols, flavones, flavanols, anthocyanidin, chalcones, and isoflavones). These substances are the result of secondary plant metabolism and are distinguished by the presence of several phenolic groups connected to complicated structures.

Unlike the basic metabolites (chlorophyll, amino acids, and simple carbohydrates), which mediate in the processes of absorption, respiration, transport, and differentiation of plants, the secondary metabolites have significant ecological roles [67].

Honey's phenolic composition is mostly determined by its floral origin; in fact, this characteristic can be utilized to classify and authenticate honey, particularly in the case of unifloral types.

Honey's capacity to scavenge free radicals and transform them into less harmful, more stable molecules is what gives it its antioxidant properties. By releasing hydrogen from their hydroxyl groups, phenolic compounds neutralize free radicals; the activity of the chemical is determined by the amount of groups [68].

The majority of the naturally occurring low-molecular-weight flavonoids are soluble in water. They are composed of two benzene rings that are alternated with a linear carbon chain consisting of three atoms (C6-C3-C6). This structure frequently reorganizes to form three rings, A, B, and C, each containing fifteen carbon atoms. When flavonoids are not linked to sugars, they are referred to as aglycones. Flavonoids are divided into groups based on how much the C ring has been oxidized: isoavones, anthocyanins, anthocyanidins, flavonols, flavones, and flavonones. Among them, flavones, flavonols, and flavonols are the most prevalent in honey [69]

A phenolic ring and at least one organic carboxylic acid function are characteristics of phenolic acids, also known as phenol carboxylic acids. These acids are classified into three groups based on their structure: C6-C3 (p-coumaric, ferulic, and caffeic acid), C6-C2 (acetophenones and phenylacetic acids), and C6-C1 (syringic, vanillic, and gallic acid). The majority of these substances are typically attached to cellulose and lignin, which are the plant's structural components, as well as to other organic molecules like glucose, other sugars, or flavonoids [31].

Honey's phenolic content varies both qualitatively and quantitatively depending on the source and botanical origin of the honey nectar, and the phenolic compounds present in honey can be a determining factor in the manufacture of certain honeys (like chestnut).

Moreover, phenolic compounds have been linked to honey's health benefits, such as its antimicrobial, antioxidant, and anti-inflammatory properties. Phenolic compounds can also act as marker compounds to help distinguish between different types of honey.

Numerous studies have found a correlation between honey's phenolic content and antioxidant activity [18]. The phenolic profile of honey also reveals the type of bees, their botanical origin, the season, and the area. In one study, the phenolic content of honey from stingless bees and common honeybees is compared. In comparison to Tualang honey, which has a total phenolic content of 183 mg GAE (gallic acid equivalent), Malaysian stingless bee honey has a total phenolic content of around 235 mg, while its flavonoid concentration is 100 mg CE (catechin equivalent). The antioxidant content of stingless bee honey was found to be substantially higher than that of regular honey in the same study [70].

The primary medicinal properties of honey are ascribed to its polyphenol content, since these compounds are the most prevalent phytochemicals and have drawn significant interest from scientific communities as potential preventive agents against degenerative and chronic inflammatory diseases [71]. The bioaccessibility and bioavailability of polyphenols in the human body regulate these functions. Polyphenols that are broken down mechanically and biochemically in the small intestine or liberated from the food matrix have the potential to be both bioavailable and bioactive in terms of bioaccessibility.

Nevertheless, bioaccessibility may be altered or increased depending on how polyphenols interact molecularly with other food ingredients throughout the digestive process [72]. The release of polyphenols for gastrointestinal tract absorption is largely dependent on the presence of dietary fibers in the meal matrix [73]. Less than 10% of polyphenols (aglycones and glucoside conjugate) are absorbed in the upper gastrointestinal tract as a result of the influence of food matrix, according to a small number of clinical trials, while the remaining phenolic compounds go through microfloral metabolization, at which point bioactive metabolites cross the colonic mucosa to plasma [74].

It is also evident that polyphenols may have a strong affinity for fiber and protein, particularly when all of the aforementioned elements are present together. It is therefore possible that the presence of polyphenol in a food matrix including minuscule levels of dietary fiber, protein, and fats will decrease molecular interactions and raise the ratio of bioaccessibility to bioavailability from honey.

Regarding this, manuka honey did not change in terms of its phenolic content or antioxidant activity when compared to other commercial honey samples during the in vitro digestion process [75]. Furthermore, post- and pre-digested manuka honey (ranging from 1–3 mg/mL) had a strong protective effect against hydrogen peroxide-induced DNA damage in the Caco-2 cell line. This shows that manuka honey's polyphenols are not impacted by any molecular alterations brought on by the digestive process, possibly contributing to their high bioaccessibility ratio [75].

The most common polyphenols in honey are shown in Table 1 and then discussed in text.

Table 1. The most common polyphenols found in honey, also discussed in the following text.

Compound	Chemical Structure	Mechanism of Action
Apigenin	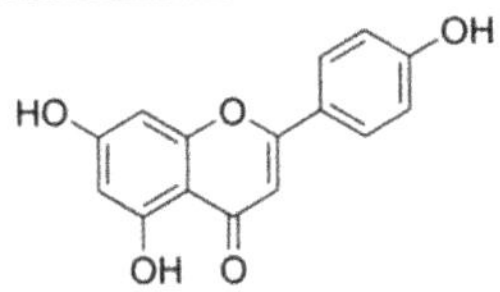	Apigening triggers Wnt/b-catenin signaling, promoting osteogenic differentiation of mesenchymal stem cell. Apigenin also boosted the expression of Runx2 and Osterix to promote bone formation and improved osteogenic differentiation [76].
Chrysin	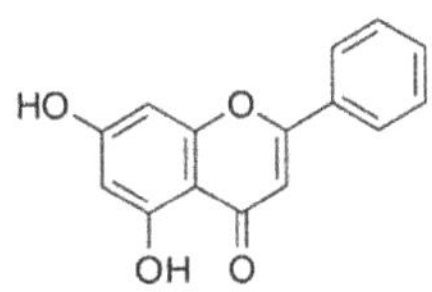	Chyrsin can encourage osteoblast differentiation and mesenchymal stem cell proliferation by activating the ERK/MAPK pathway [77].
Galangin	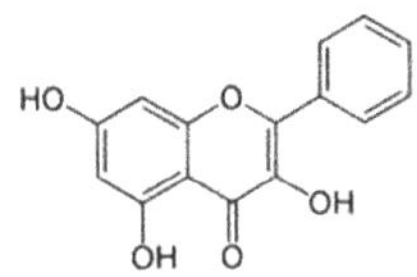	Galangin possesses a number of biological characteristics, including the capacity to alter metabolic enzymes and anti-oxidative, anti-mutagenic, anti-clastogenic, and radical scavenging qualities [78].
Genistein	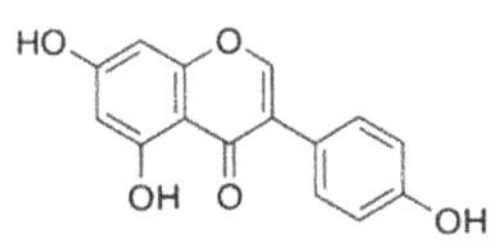	Genistein can protect cells from oxidative stress and cellular damage by triggering the cellular signaling pathway of Nrf2 (Nuclear factor erythroid 2-related factor 2) [79].
Luteolin	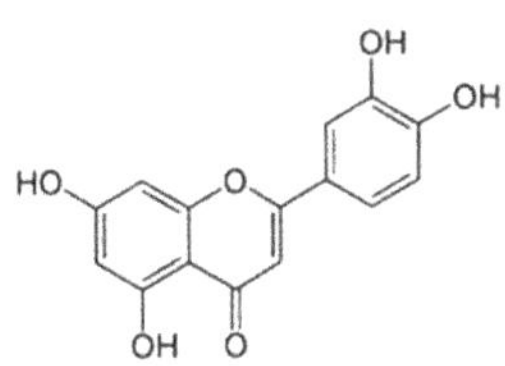	Luteolin can promote astrocyte differentiation and increase astrogenesis in human pluripotent stem cells (hPSCs) via the Wnt-β-catenin-BMP2-STAT3 signaling pathway. By deactivating NF-κB and upregulating Nrf2, systemic luteolin therapy enhances wound healing by lowering inflammation and oxidative stress [80,81].
Myricetin	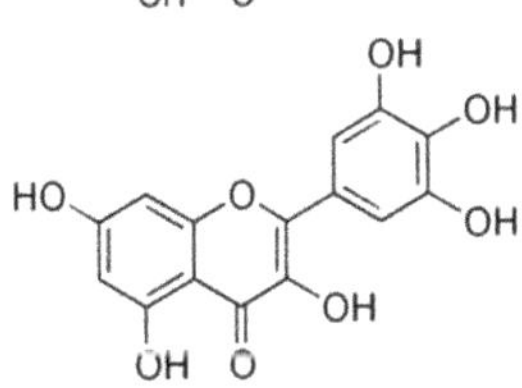	Myricetin enhances the osteogenic development of human periodontal ligament stem cells (hPDLSCs) by more efficiently regulating the BMP-2/Smad and ERK/JNK/MAPK signaling pathways [31,82].
Pinobanksin	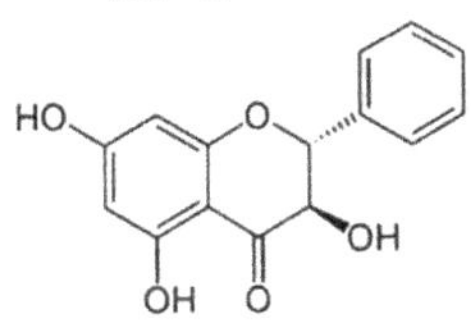	The biological effects of pinobanksin are antioxidant, anti-inflammatory, and antibacterial properties [83].
Pinocembrin	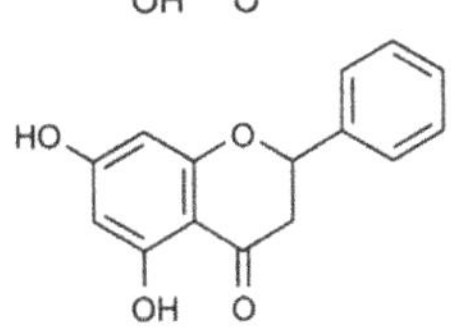	By triggering the PI3K-eNOS-NO signaling pathway, pinocembrin can improve the biological activity of endothelial progenitor cells (EPCs) in the bone marrow [84].

Table 1. *Cont.*

Compound	Chemical Structure	Mechanism of Action
ng Quercetin	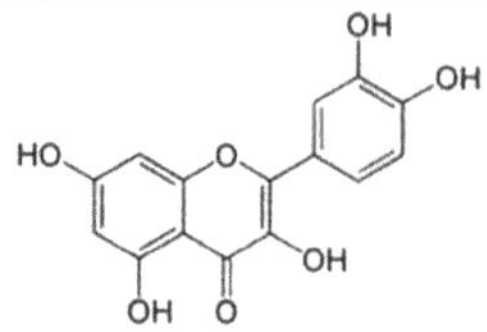	The ERK and p38 signaling pathways allow quercetin to stimulate bone marrow stem cell (BMSC) osteogenic differentiation and proliferation [85].
Caffeic acid	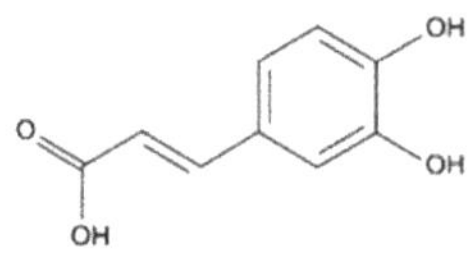	Caffeine has anti-inflammatory, anticancer, and antioxidant properties [86,87].
Chlorogenic acid	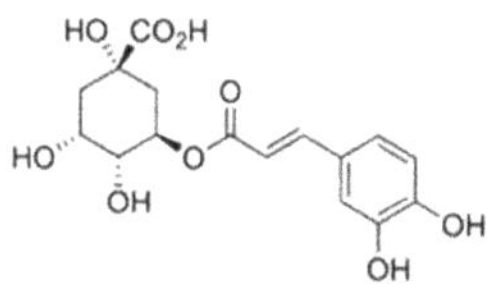	Chlorogenic acid exhibits antioxidant, anti-inflammatory, and anti-apoptotic properties [88].

3.6.1. Apigenin

Apigenin (4,5,7-trihydroxyflavone) is a low-toxicity flavonoid found abundantly in plants and has been extensively explored. Apigenin is mostly found in fruits, vegetables, nuts, herbs, tea, and honey.

The primary sources of this chemical are plants in the Asteraceae family, specifically those in the genera *Artemisia* [89], *Achillea* [90,91], *Matricaria* [92], and *Tanacetum* [93]. Numerous pieces of research conducted over the years have suggested that apigenin possesses a wide range of intriguing pharmacological properties and prospective applications in nutraceuticals. Honey contains approximately 0.03 mg/100 g of this chemical [16].

Some authors have reported that apigenin, by triggering Wnt/b-catenin signaling, promotes osteogenic differentiation of mesenchymal stem cells and quickens the repair of bone fractures [76]. According to Zhang et al., apigenin boosted the expression of Runx2 and Osterix to promote bone formation and improved the osteogenic differentiation of human mesenchymal stem cells via JNK and p38 MAPK pathways [94].

Moreover, apigenin anticancer effects in a variety of cell lines have been extensively documented. It demonstrated an anti-proliferative effect on liver, breast, colon, neuroblastoma, and cervical cancer cell lines [95].

3.6.2. Chrysin

Chrysin (5,7-dihydroxy-2-phenyl-4H-chromen-4-one) is a flavonoid that naturally occurs in the flavone class of polyphenolic chemicals with a 15-carbon structure. Chrysin is mostly found in nature in passion fruit (*Passiflora* sp.), honey, and propolis [96–98].

Many varieties of honey have elevated quantities of chrysin. Forest honey has a chrysin concentration of 0.53 mg/100 g [99].

By stimulating the ERK/MAPK pathway, chyrsin can promote osteoblastic differentiation and mesenchymal stem cell proliferation [100]. Moreover, chrysin can stimulate many signaling pathways that are involved in the development of dental plug stem cells into osteogenic or odontogenic forms. Osteogenic differentiation is produced by increased alkaline phosphatase (ALP) activity and modified expression of osteocalcin, Runx2, and collagen type 1 by chrysin [77]. Chrysin causes osteogenic differentiation of dental pulp stem cells through the activation of the Smad3 signaling pathway, a crucial part of the TGF-β signaling system [77].

3.6.3. Galangin

Honey naturally contains the flavonoid galangin (3,5,7-trihydroxy-2-phenylchromen-4-one, also known as 3,5,7-trihydroxyflavone). According to Habryka et al. [101], honey shows a galangin concentration of about 0.334 mg/100 g. High amounts of galangin can be found in honey and *Alpinia officinarum*, a plant that has long been used in Asia as a spice and herbal remedy for a wide range of illnesses. Galangin has several biological properties, such as the ability to modulate metabolic enzymes, and has anti-mutagenic, anti-clastogenic, anti-oxidative, and radical scavenging properties [78].

3.6.4. Genistein

Numerous dietary legumes, including fava beans, lupine, soybeans, green lentils, alfalfa, and peas, naturally contain the chemical genistein, a 7-hydroxyisoflavone. As an anti-inflammatory, antioxidant, anti-Alzheimer, antidiabetic, or anticancer drug, genistein is used to treat a wide range of illnesses. Furthermore, through activating the Nrf2 (Nuclear factor erythroid 2-related factor 2) cellular signaling pathway, genistein can shield cells from oxidative stress and cellular toxicity [79].

3.6.5. Luteolin

Luteolin (3′,4′,5,7-tetrahydroxyflavone) is a flavone found in celery, parsley, green pepper, and chamomile tea. It possesses anti-inflammatory and antioxidant characteristics.

According to Hossen et al., honey contains around 0.28 mg of luteolin per 100 g. Numerous biological actions, including anti-inflammatory, antibacterial, antioxidant, and anticancer properties, are known to be exhibited by luteolin [16].

Through the Wnt-β-catenin-BMP2-STAT3 signaling pathway, the administration of the natural flavone luteolin can stimulate astrocyte differentiation and boost astrogenesis in human pluripotent stem cells (hPSCs). Luteolin also stimulates the Wnt/β-catenin signaling pathway to hasten the osteogenic development of human periodontal ligament cells. In type 1 diabetic cardiomyopathy, luteolin reduces oxidative stress and thereby improves outcomes following heart failure [80]. Lutein can also help the wound healing mechanism since it promotes fibroblast proliferation. Systemic luteolin treatment improves wound healing by reducing inflammation and oxidative stress by inactivating NF-κB and upregulating Nrf2 [81].

3.6.6. Myricetin

Chemically known as 3,5,7-trihydroxy-2-(3,4,5-trihydroxyphenyl) chromen-4-one, myricetin is a naturally occurring polyhydroxy flavonoid substance that is extracted from *Myrica rubra* leaves and bark.

Berries, tea, nuts, fruits, and vegetables are the most prevalent sources of myricetin. The amounts in black fruits range from 14 to 142 mg/kg. The most prevalent flavonoid in black currants is myricetin, and the amount of this flavonoid varies greatly between varieties. In addition, myricetin can be found in honey. The range of myricetin is around 29.2–289.0 μg/100 g of honey [102].

In cancer cell lines, myricetin causes apoptosis through its antiviral, anti-carcinogenic, anti-inflammatory, and antioxidant properties. Additionally, thioredoxin reductase activity in A549 lung cancer cells and matrix metalloproteinase 2 (MMP-2) enzyme activity in colorectal carcinoma cells were both suppressed by myricetin [31].

Moreover, myricetin regulates the BMP-2/Smad and ERK/JNK/MAPK signaling pathways more effectively, which improves the osteogenic differentiation of human periodontal ligament stem cells (hPDLSCs) [82].

3.6.7. Pinobanksin

Pinobanksin (3,5,7-trihydroxy-2-phenyl-chroman-4-one) is a phenyl allyl flavanone that is frequently found in fruits, propolis, honey, bee bread, bee pollen, and bee wax in addition to the heartwood of the *Pinus* genus. Recent research has revealed pinobanksin's biological actions, including its strong antioxidant, anti-inflammatory, and antimicrobial qualities, marking it as promising pharmaceutical.

Since pinobanksin may also be found in many types of propolis and honey that is sourced from across the globe, it can also be referred to as a dietary flavonoid [83]. Pinobanksin is known for its antioxidant characteristics, working by preventing the peroxidation of low-density lipoprotein and by decreasing alpha-tocopherol radicals through its electron-donor properties.

3.6.8. Pinocembrin

One of the principal flavonoids present in a wide variety of plants, fungi, and bee products, primarily propolis and honey, is pinocembrin (5,7-dihydroxyflavone).

Numerous in vitro and preclinical investigations have demonstrated the pharmacological properties of pinocembrin, including anticancer, neuroprotective, anti-inflammatory, antibacterial, and antioxidant properties [103]. According to Jasicka-Misiak et al. [104], this molecule has a concentration of 1.60–1.85 mg/100 g in honey. Pinocembrin, which has been found in propolis and honey, most likely originated in indigenous plant pollen grains and was carried by bees [105].

Pinocembrin can also enhance the biological function of endothelial progenitor cells (EPCs) in the bone marrow via activating the PI3K-eNOS-NO signaling pathway. As crucial transcription factors in osteoblast formation, Runx2 and Osterix, as well as their enhancers, Dlx5 and Msx2, had higher mRNA expression when exposed to pinocembrin [84].

3.6.9. Quercetin

One natural flavonoid that can be found in food, including honey, is quercetin. According to Yayinie et al. [106], honey has a quercetin concentration of 1.67–5.08 mg/100 g.

Quercetin has numerous positive health impacts, including preventing diseases like osteoporosis, lung cancer, and cardiovascular disease. Through the ERK and p38 signaling pathways, quercetin may promote the in vitro osteogenic differentiation and proliferation of bone marrow stem cells (BMSCs) [85]. Numerous investigations have also revealed that quercetin promotes the differentiation of osteoblasts into BMSCs [107].

Since quercetin is one of the most prevalent compounds in plants, honeybee diets are high in it. It is currently unknown to what degree quercetin consumption will lower the content of ingested poisons, although it can support the detoxifying system in honeybees. In honeybees, consumption of quercetin-sucrose paste (10 mg quercetin/g) might decrease fluvalinate's acute toxicity and prolong bees' lifespans when pyrethroids are present [108].

3.6.10. Caffeic Acid

Caffeic acid is a phenolic acid that is mostly found in human diets and is formed through the secondary metabolism of vegetables. As a result of its ability to prevent the growth of bacteria, fungi, and insects, caffeic acid helps plants defend themselves against diseases, predators, and other threats.

Caffeic acid is present in honey, and, in particular, has been found in buckwheat (0.54 mg/kg), multifloral (0.23 mg/kg), and linden (0.15 mg/kg) honeys [109]. Caffeic acid has been shown to exhibit antioxidant, anti-inflammatory, and anticancer activities.

3.6.11. Chlorogenic Acid

Another common phenolic acid of honey is chlorogenic acid (3-(3,4-dihydroxycinnamoyl) quinic acid). Lu et al. [110] report that it is a common ingredient in fruits and vegetables, as well as the second primary component of coffee beans. Chlorogenic acid exhibits anti-inflammatory (through GPR37 activation), antioxidant, and anti-apoptotic activities [86–88,111]. This chemical is present in honey at a concentration of roughly 0.17 mg/100 g [112].

4. Biological Qualities of Honey

4.1. Honey Application

While some modern chemicals have failed to treat wounds, honey is the oldest known wound-healing agent. Its antibacterial, antiviral, anti-inflammatory, and antioxidant properties have led to additional experimental study demonstrating its effectiveness in wound healing. According to research, honey can be used to treat and manage acute wounds as well as mild to moderate superficial and partial-thickness burns [80]. More research is required to support the existing evidence, even though some studies have shown that honey is effective in treating wounds and leg ulcers. In particular, Ranzato and colleagues [5] have demonstrated the role, in the wound-healing effects boosted by honey, of epithelial-to-mesenchymal transition activation in keratinocytes and the pivotal role of hydrogen peroxide. In particular, hydrogen peroxide is able to cross the plasma membrane of a cell via a specific transporter (i.e., aquaporin-3, [22]) and then initiates a Ca^{2+}-signaling cascade, starting the wound closure process [7,113].

Strong evidence supports honey's potential benefits in the management of diabetes mellitus [114]. These findings highlight the potential benefits of utilizing honey or other strong antioxidants in addition to prescription antidiabetic medications to treat diabetes mellitus. Nonetheless, a number of uncertainties remain, particularly in relation to the possibility of managing diabetes mellitus through therapies that address both hyperglycemia and oxidative stress. Additionally, honey may have therapeutic benefits for managing diabetes that go beyond only lowering blood sugar levels; honey may also help treat related metabolic complications [114].

According to recent research, honey shows anticancer properties through a number of different routes [115]. According to studies, honey has anticancer properties by interfering with several cell-signaling pathways, such as those that trigger apoptosis [116,117]. Furthermore, honey may be able to prevent a number of tumor types in animal models. Nevertheless, additional research is required to enhance our comprehension of honey's beneficial effects on cancer [118].

Flavonoids and phenolics, two types of antioxidants found in honey, may lower the risk of cardiovascular failure [16,119]. Flavonoids' preventive properties, including their antithrombotic, anti-ischemic, antioxidant, and vasorelaxant properties, lower the risk of coronary heart disease. Numerous studies have demonstrated the potential pharmacological role of certain honey polyphenols in lowering cardiovascular diseases [120]. To further validate these molecules in medical applications, however, in vitro and in vivo research as well as clinical trials should be undertaken.

It has been suggested that numerous gastrointestinal illnesses, including periodontal disease and other oral problems, dyspepsia, and oral rehydration therapy, could benefit from honey [121]. Although a clinical trial of manuka honey therapy to promote *Helicobacter pylori* eradication failed to show an effective treatment, in vitro investigations suggest that honey has bactericidal action against *Helicobacter* [122,123].

The advancement of nanotechnology has made it possible for humans to better utilize the potential of numerous natural substances. However, traditional nanotechnology frequently employs harsh synthesis conditions along with hazardous and environmentally damaging chemical substances. Nevertheless, "green chemistry" methods are transforming the potential for creating cost-effective and ecologically friendly technology, even for tissue engineering. Honey appears to be one of the most promising natural substances with which researchers are implementing this new "green" strategy. Moreover, because 3D honey-based porous scaffolds can replicate the architecture of a native microenvironment and may allow for the controlled release of honey at the site of injury for rapid regeneration, they are gaining more interest in the fields of biomedicine and wound healing [124].

4.2. Quality of Honey

Natural bee honey is a product produced by honeybees (*Apis mellifera* L.), which use plant nectar as well as excreta from insects that extract plant juice or secretions from living plant parts. These are then mixed with particular bee secretions, kept in honeycombs, evaporated, and allowed to mature there.

The grade of bee honey should be high enough for human consumption. The following parameters are required in each country: variety determination, water content, pH, insoluble impurities (no more than 0.1 g/100 g), proline (no less than 25 mg/100 g), diastase activity (general, no less than eight on the Schade scale), electrical conductivity, free acidity (general, no more than 50 milli-equivalents acid/1000 g), and water content. Color (pFund value) and total phenolic content (TPC) are additional factors. Standard techniques, such as spectrophotometric, refractometric, titration, and melissopalynological approaches, are employed to evaluate the quality of honey [125].

Importing honey has become necessary as a result of the rise in consumer demand for bee honey in recent years due to its health-promoting qualities. Adding less expensive high-fructose corn syrup (HFCS) is one method of faking bee honey. One of the most commonly faked foodstuffs is bee honey [126]. By using the $^{13}C/^{12}C$ isotope ratio, Çinar et al. [127] looked for a way to identify bee honey that has been adulterated with HFCS.

Despite the fact that honey's quality standards are well-established globally, unfortunately none of them take into account the biological characteristics or health benefits of honey [128].

5. Conclusions

As a naturally occurring substance, honey contains important minerals, vitamins, proteins, enzymes, amino acids, and bioactive substances including phenolic compounds, which have fascinating biological qualities both in vitro and in vivo.

For instance, by altering a number of molecular pathways, honey can prevent the growth of cancerous cells and effectively combat microbial infections. It also has protective effects on the neurological, respiratory, and gastrointestinal systems.

Researchers have verified that, when compared to other functional foods, honey polyphenols have a high bioavailability value. Nevertheless, a complete picture of the molecular mechanisms of polyphenols abundant in honey is lacking. To completely understand the genome-wide effects of honey and patterns of global gene expression, protein translation, intracellular signaling pathways, and metabolite formation in response to specific chemicals, more research is therefore still required in nutrigenomics analysis.

In conclusion, investigating how honey affects gene expression profiling, focusing especially on human intervention trials (ideally large-scale randomized placebo-controlled studies), will help to develop effective strategies for mitigating chronic inflammatory diseases and provide valuable insights into the prophylactic and therapeutic uses of honey.

Therefore, to encourage the general public to use this nutritious food, to support a healthy lifestyle, and to prevent some of the most frequent pathologies, a greater understanding of the mechanisms and contributing elements of the honey effect will be essential.

Author Contributions: Conceptualization, E.R. and S.M.; methodology, E.R.; software, G.B.; validation, G.B., S.M. and E.R.; resources, S.M.; data curation, G.B. and E.R.; writing—original draft preparation, G.B.; writing—review and editing, E.R.; visualization, E.R.; supervision, S.M.; project administration, S.M.; funding acquisition, S.M. All authors have read and agreed to the published version of the manuscript.

Funding: Project funding from NextGeneration EU and Compagnia di San Paolo.

Institutional Review Board Statement: Not applicable.

Informed Consent Statement: Not applicable.

Data Availability Statement: No new data were created or analyzed in this study. Data sharing is not applicable to this article.

Acknowledgments: We thank our beekeeper friends, who always encourage us to pursue our work on honey.

Conflicts of Interest: The authors declare no conflicts of interest.

References

1. Martinotti, S.; Ranzato, E. Honey's healing history. In *Cellular and Molecular Mechanisms of Honey Wound Healing*; Ranzato, E., Martinotti, S., Eds.; Nova Publishers Inc.: Hauppage, NY, USA, 2014.
2. Erler, S.; Moritz, R.F. Pharmacophagy and pharmacophory: Mechanisms of self-medication and disease prevention in the honeybee colony (*Apis mellifera*). *Apidologie* **2016**, *47*, 389–411. [CrossRef]
3. Carter, D.A.; Blair, S.E.; Cokcetin, N.N.; Bouzo, D.; Brooks, P.; Schothauer, R.; Harry, E.J. Therapeutic Manuka Honey: No Longer So Alternative. *Front. Microbiol.* **2016**, *7*, 569. [CrossRef] [PubMed]
4. Oryan, A.; Alemzadeh, E.; Moshiri, A. Biological properties and therapeutic activities of honey in wound healing: A narrative review and meta-analysis. *J. Tissue Viability* **2016**, *25*, 98–118. [CrossRef]
5. Ranzato, E.; Martinotti, S.; Burlando, B. Epithelial mesenchymal transition traits in honey-driven keratinocyte wound healing: Comparison among different honeys. *Wound Repair. Regen.* **2012**, *20*, 778–785. [CrossRef] [PubMed]
6. Martinotti, S.; Ranzato, E. Honey, Wound Repair and Regenerative Medicine. *J. Funct. Biomater.* **2018**, *9*, 34. [CrossRef] [PubMed]
7. Martinotti, S.; Bonsignore, G.; Ranzato, E. Applications of Beehive Products for Wound Repair and Skin Care. *Cosmetics* **2023**, *10*, 127. [CrossRef]
8. Martinotti, S.; Calabrese, G.; Ranzato, E. Honeydew honey: Biological effects on skin cells. *Mol. Cell Biochem.* **2017**, *435*, 185–192. [CrossRef]
9. Bucekova, M.; Sojka, M.; Valachova, I.; Martinotti, S.; Ranzato, E.; Szep, Z.; Majtan, V.; Klaudiny, J.; Majtan, J. Bee-derived antibacterial peptide, defensin-1, promotes wound re-epithelialisation in vitro and in vivo. *Sci. Rep.* **2017**, *7*, 7340. [CrossRef]
10. Bucekova, M.; Juricova, V.; Monton, E.; Martinotti, S.; Ranzato, E.; Majtan, J. Microwave processing of honey negatively affects honey antibacterial activity by inactivation of bee-derived glucose oxidase and defensin-1. *Food Chem.* **2018**, *240*, 1131–1136. [CrossRef]
11. Martinotti, S.; Bucekova, M.; Majtan, J.; Ranzato, E. Honey: An Effective Regenerative Medicine Product in Wound Management. *Curr. Med. Chem.* **2019**, *26*, 5230–5240. [CrossRef]
12. Martinotti, S.; Bonsignore, G.; Patrone, M.; Ranzato, E. Correlation between Honey Parameters and Wound Healing Properties: The Case of Piedmont (Italy) Samples. *Curr. Pharm. Biotechnol.* **2024**. [CrossRef] [PubMed]
13. Nweze, J.A.; Okafor, J.I.; Nweze, E.I.; Nweze, J.E. Comparison of antimicrobial potential of honey samples from *Apis mellifera* and two stingless bees from Nsukka, Nigeria. *J. Pharmacogn. Nat. Prod.* **2016**, *2*, 1–7. [CrossRef]
14. Alam, F.; Islam, M.A.; Gan, S.H.; Khalil, M.I. Honey: A potential therapeutic agent for managing diabetic wounds. *Evid. Based Complement. Alternat Med.* **2014**, *2014*, 169130. [CrossRef]
15. Al-Waili, N.; Salom, K.; Al-Ghamdi, A.A. Honey for wound healing, ulcers, and burns; data supporting its use in clinical practice. *Sci. World J.* **2011**, *11*, 766–787. [CrossRef]
16. Hossen, M.S.; Ali, M.Y.; Jahurul, M.H.A.; Abdel-Daim, M.M.; Gan, S.H.; Khalil, M.I. Beneficial roles of honey polyphenols against some human degenerative diseases: A review. *Pharmacol. Rep.* **2017**, *69*, 1194–1205. [CrossRef]
17. Meo, S.A.; Al-Asiri, S.A.; Mahesar, A.L.; Ansari, M.J. Role of honey in modern medicine. *Saudi J. Biol. Sci.* **2017**, *24*, 975–978. [CrossRef] [PubMed]
18. Abd Jalil, M.A.; Kasmuri, A.R.; Hadi, H. Stingless Bee Honey, the Natural Wound Healer: A Review. *Skin Pharmacol. Physiol.* **2017**, *30*, 66–75. [CrossRef]
19. Israili, Z.H. Antimicrobial properties of honey. *Am. J. Ther.* **2014**, *21*, 304–323. [CrossRef]
20. Alvarez-Suarez, J.M.; Gasparrini, M.; Forbes-Hernández, T.Y.; Mazzoni, L.; Giampieri, F. The Composition and Biological Activity of Honey: A Focus on Manuka Honey. *Foods* **2014**, *3*, 420–432. [CrossRef]
21. Almasaudi, S.B.; Al-Nahari, A.A.M.; Abd El-Ghany, E.S.M.; Barbour, E.; Al Muhayawi, S.M.; Al-Jaouni, S.; Azhar, E.; Qari, M.; Qari, Y.A.; Harakeh, S. Antimicrobial effect of different types of honey on. *Saudi J. Biol. Sci.* **2017**, *24*, 1255–1261. [CrossRef] [PubMed]
22. Martinotti, S.; Laforenza, U.; Patrone, M.; Moccia, F.; Ranzato, E. Honey-Mediated Wound Healing: H_2O_2 Entry through AQP3 Determines Extracellular Ca. *Int. J. Mol. Sci.* **2019**, *20*, 764. [CrossRef]
23. Fahim, H.; Dasti, J.I.; Ali, I.; Ahmed, S.; Nadeem, M. Physico-chemical analysis and antimicrobial potential of Apis dorsata, *Apis mellifera* and *Ziziphus jujube* honey samples from Pakistan. *Asian Pac. J. Trop. Biomed.* **2014**, *4*, 633–641. [CrossRef]
24. Molan, P.; Rhodes, T. Honey: A Biologic Wound Dressing. *Wounds* **2015**, *27*, 141–151.
25. Purbafrani, A.; Ghazizade Hashemi, S.A.; Bayyenat, S.; Moghaddam, H.T.; Saeidi, M. The benefits of honey in holy Quran. *Int. J. Pediatr.* **2014**, *2*, 67–73.
26. Aziz, M.; Jacob, A.; Yang, W.L.; Matsuda, A.; Wang, P. Current trends in inflammatory and immunomodulatory mediators in sepsis. *J. Leukoc. Biol.* **2013**, *93*, 329–342. [CrossRef]
27. Viuda-Martos, M.; Ruiz-Navajas, Y.; Fernández-López, J.; Pérez-Alvarez, J.A. Functional properties of honey, propolis, and royal jelly. *J. Food Sci.* **2008**, *73*, R117–R124. [CrossRef]
28. Hadagali, M.D.; Chua, L.S. The anti-inflammatory and wound healing properties of honey. *Eur. Food Res. Technol.* **2014**, *239*, 1003–1014. [CrossRef]

29. Yuksel, P.; Saribas, S.; Bagdatli, Y. Comparison of the VersaTrek and BACTEC MGIT 960 systems for the contamination rate, time of detection and recovery of mycobacteria from clinical specimens. *Afr. J. Microbiol. Res.* **2011**, *5*, 985–989.
30. Mesaik, M.A.; Dastagir, N.; Uddin, N.; Rehman, K.; Azim, M.K. Characterization of immunomodulatory activities of honey glycoproteins and glycopeptides. *J. Agric. Food Chem.* **2015**, *63*, 177–184. [CrossRef]
31. Alvarez-Suarez, J.M.; Giampieri, F.; Battino, M. Honey as a source of dietary antioxidants: Structures, bioavailability and evidence of protective effects against human chronic diseases. *Curr. Med. Chem.* **2013**, *20*, 621–638. [CrossRef]
32. Rao, P.V.; Krishnan, K.T.; Salleh, N.; Gan, S.H. Biological and therapeutic effects of honey produced by honey bees and stingless bees: A comparative review. *Rev. Bras. De Farmacogn.* **2016**, *26*, 657–664. [CrossRef]
33. Majtan, J. Honey: An immunomodulator in wound healing. *Wound Repair. Regen.* **2014**, *22*, 187–192. [CrossRef] [PubMed]
34. Osés, S.M.; Pascual-Maté, A.; Fernández-Muiño, M.A.; López-Díaz, T.M.; Sancho, M.T. Bioactive properties of honey with propolis. *Food Chem.* **2016**, *196*, 1215–1223. [CrossRef] [PubMed]
35. Moniruzzaman, M.; Sulaiman, S.A.; Khalil, M.I.; Gan, S.H. Evaluation of physicochemical and antioxidant properties of sourwood and other Malaysian honeys: A comparison with manuka honey. *Chem. Cent. J.* **2013**, *7*, 138. [CrossRef] [PubMed]
36. Chaikham, P.; Prangthip, P. Alteration of antioxidative properties of longan flower-honey after high pressure, ultra-sonic and thermal processing. *Food Biosci.* **2015**, *10*, 1–7. [CrossRef]
37. Escuredo, O.; Míguez, M.; Fernández-González, M.; Carmen Seijo, M. Nutritional value and antioxidant activity of honeys produced in a European Atlantic area. *Food Chem.* **2013**, *138*, 851–856. [CrossRef]
38. Alqarni, A.S.; Owayss, A.A.; Mahmoud, A.A. Mineral content and physical properties of local and imported honeys in Saudi Arabia. *J. Saudi Chem. Soc.* **2012**, *5*, 618–625. [CrossRef]
39. Ciulu, M.; Solinas, S.; Floris, I.; Panzanelli, A.; Pilo, M.I.; Piu, P.C.; Spano, N.; Sanna, G. RP-HPLC determination of water-soluble vitamins in honey. *Talanta* **2011**, *83*, 924–929. [CrossRef]
40. Pontes, M.; Marques, J.C.; Câmara, J.S. Screening of volatile composition from Portuguese multifloral honeys using headspace solid-phase microextraction-gas chromatography-quadrupole mass spectrometry. *Talanta* **2007**, *74*, 91–103. [CrossRef]
41. Kamal, M.A.; Klein, P. Determination of sugars in honey by liquid chromatography. *Saudi J. Biol. Sci.* **2011**, *18*, 17–21. [CrossRef]
42. Escuredo, O.; Dobre, I.; Fernández-González, M.; Seijo, M.C. Contribution of botanical origin and sugar composition of honeys on the crystallization phenomenon. *Food Chem.* **2014**, *149*, 84–90. [CrossRef]
43. Tornuk, F.; Karaman, S.; Ozturk, I.; Toker, O.S.; Tastemur, B.; Sagdic, O.; Dogan, M.; Kayacier, A. Quality characterization of artisanal and retail Turkish blossom honeys: Determination of physicochemical, microbiological, bioactive properties and aroma profile. *Ind. Crops Prod.* **2013**, *46*, 124–131. [CrossRef]
44. Kaskoniene, V.; Venskutonis, P.R.; Ceksteryte, V. Composition of volatile compounds of honey of various floral origin and beebread collected in Lithuania. *Food Chem.* **2008**, *111*, 988–997. [CrossRef]
45. Chernetsova, E.S.; Morlock, G.E. Assessing the capabilities of direct analysis in real time mass spectrometry for 5-hydroxymethylfurfural quantitation in honey. *Int. J. Mass Spectrom.* **2012**, *314*, 22–32. [CrossRef]
46. Moreira, R.F.A.; Maria, C.A.B.; Pietroluongo, M.; Trugo, L.C. Chemical changes in the volatile fractions of Brazilian honeys during storage under tropical conditions. *Food Chem.* **2010**, *121*, 697–704. [CrossRef]
47. Wang, S.; Guo, Q.; Wang, L.; Lin, L.; Shi, H.; Cao, H.; Cao, B. Detection of honey adulteration with starch syrup by high performance liquid chromatography. *Food Chem.* **2015**, *172*, 669–674. [CrossRef]
48. Ota, M.; Kohmura, M.; Kawaguchi, H. Characterization of a new Maillard type reaction product generated by heating 1-deoxymaltulosyl-glycine in the presence of cysteine. *J. Agric. Food Chem.* **2006**, *54*, 5127–5131. [CrossRef]
49. Hermosín, I.; Chicón, R.M.; Cabezudo, M.D. Free amino acid composition and botanical origin of honey. *Food Chem.* **2003**, *83*, 263–268. [CrossRef]
50. Won, S.A.; Li, C.; Kim, J.; Rhee, H. Immunological characterization of honeymajor protein and its application. *Food Chem.* **2009**, *113*, 1334–1338. [CrossRef]
51. Sak-Bosnar, M.; Sakač, N. Direct potentiometric determination of diastase activity in honey. *Food Chem.* **2012**, *135*, 827–831. [CrossRef]
52. Iglesias, M.T.; Martín-Alvarez, P.J.; Polo, M.C.; de Lorenzo, C.; Gonzalez, M.; Pueyo, E. Changes in the free amino acid contents of honeys during storage at ambient temperature. *J. Agric. Food Chem.* **2006**, *54*, 9099–9104. [CrossRef] [PubMed]
53. Truzzi, C.; Annibaldi, A.; Illuminati, S.; Finale, C.; Scarponi, G. Determination of proline in honey: Comparison between official methods, optimization and validation of the analytical methodology. *Food Chem.* **2014**, *150*, 477–481. [CrossRef] [PubMed]
54. Manzanares, A.B.; García, H.; Galdón, B.R.; Rodríguez, E.R.; Romero, C.D. Physicochemical characteristics of minor monofloral honeys from Tenerife, Spain. *Food Sci. Technol.* **2014**, *55*, 572–578.
55. Moreira, R.F.A.; Maria, C.A.B.; Pietroluongo, M.; Trugo, L.C. Chemical changes in the non-volatile fraction of Brazilian honeys during storage under tropical conditions. *Food Chem.* **2007**, *104*, 1236–1241. [CrossRef]
56. Czipa, N.; Borbély, M.; Gyori, Z. Proline content of different honey types. *Acta Aliment.* **2012**, *41*, 26–32. [CrossRef]
57. Mato, I.; Huidobro, J.F.; Simal-Lozano, J.; Sancho, M.T. Significance of nonaromatic organic acids in honey. *J. Food Prot.* **2003**, *66*, 2371–2376. [CrossRef]
58. Aljohar, H.I.; Maher, H.M.; Albaqami, J.; Al-Mehaizie, M.; Orfali, R.; Alrubia, S. Physical and chemical screening of honey samples available in the Saudi market: An important aspect in the authentication process and quality assessment. *Saudi Pharm. J.* **2018**, *26*, 932–942. [CrossRef]

59. Tezcan, F.; Kolayli, S.; Ulusoy, H.S.E.; Erim, F.B. Evaluation of organic acid, saccharide composition and antioxidant properties of some authentic Turkish honeys. *J. Food Nutr. Res.* **2011**, *50*, 33–40.
60. Alonso-Torre, S.R.; Huidobro, J.F.; Sancho, M.T.; Fernández-Muiño, M.A.; Cavia, M.M. Evolution of acidity of honeys from continental climates: Influence of induced granulation. *Food Chem.* **2006**, *100*, 1728–1733.
61. Cherchi, A.; Spanedda, L.; Tuberoso, C.; Cabras, P. Solid-phase extraction and high-performance liquid chromatographic determination of organic acids in honey. *J. Chromatogr. A* **1994**, *669*, 59–64. [CrossRef]
62. Nozal, M.J.; Bernal, J.L.; Gómez, L.A.; Higes, M.; Meana, A. Determination of oxalic acid and other organic acids in honey and in some anatomic structures of bees. *Apidologie* **2003**, *34*, 181–188. [CrossRef]
63. Karabagias, I.K.; Badeka, A.; Kontakos, S.; Karabournioti, S.; Kontominas, M.G. Characterisation and classification of Greek pine honeys according to their geographical origin based on volatiles, physicochemical parameters and chemometrics. *Food Chem.* **2014**, *146*, 548–557. [CrossRef] [PubMed]
64. León-Ruiz, V.; Vera, S.; González-Porto, A.V.; Andrés, M.P.S. Analysis of water-soluble vitamins in honey by isocratic RP-HPLC. *Food Anal. Methods* **2013**, *6*, 488–496. [CrossRef]
65. Madejczyk, M.; Baralkiewicz, D. Characterization of Polish rape and honeydew honey according to their mineral contents using ICP-MS and F-AAS/AES. *Anal. Chim. Acta* **2008**, *617*, 11–17. [CrossRef] [PubMed]
66. Nalda, M.J.N.; Yagu, J.L.B.; Calva, J.C.D.; Gómez, M.T.M. Classifying honeys from the Soria Province of Spain via multivariate analysis. *Anal. Bioanal. Chem.* **2005**, *382*, 311–319. [CrossRef]
67. Martinotti, S.; Calabrese, G.; Ranzato, E. Plant polyphenols and human health. In *Polyphenols: Food Sources, Bioactive Properties and Antioxidant Effects*; Cobb, D.T., Ed.; Nova Publishers Inc.: Hauppauge, NY, USA, 2014.
68. Rice-Evans, C.A.; Miller, N.J. Antioxidant activities of flavonoids as bioactive components of food. *Biochem. Soc. Trans.* **1996**, *24*, 790–795. [CrossRef]
69. Moniruzzaman, M.; Yung An, C.; Rao, P.V.; Hawlader, M.N.; Azlan, S.A.; Sulaiman, S.A.; Gan, S.H. Identification of phenolic acids and flavonoids in monofloral honey from Bangladesh by high performance liquid chromatography: Determination of antioxidant capacity. *Biomed. Res. Int.* **2014**, *2014*, 737490. [CrossRef]
70. Ranneh, Y.; Ali, F.; Zarei, M.; Akim, A.M.; Abd Hamid, H.; Khazaai, H. Malaysian stingless bee and Tualang honeys: A comparative characterization of total antioxidant capacity and phenolic profile using liquid chromatography-mass spectrometry. *LWT Food Sci. Technol.* **2018**, *89*, 1–9. [CrossRef]
71. da Silva, P.M.; Gauche, C.; Gonzaga, L.V.; Costa, A.C.; Fett, R. Honey: Chemical composition, stability and authenticity. *Food Chem.* **2016**, *196*, 309–323. [CrossRef]
72. Acosta-Estrada, B.A.; Gutiérrez-Uribe, J.A.; Serna-Saldívar, S.O. Bound phenolics in foods, a review. *Food Chem.* **2014**, *152*, 46–55. [CrossRef]
73. Bohn, T. Dietary factors affecting polyphenol bioavailability. *Nutr. Rev.* **2014**, *72*, 429–452. [CrossRef] [PubMed]
74. Erk, T.; Williamson, G.; Renouf, M.; Marmet, C.; Steiling, H.; Dionisi, F.; Barron, D.; Melcher, R.; Richling, E. Dose-dependent absorption of chlorogenic acids in the small intestine assessed by coffee consumption in ileostomists. *Mol. Nutr. Food Res.* **2012**, *56*, 1488–1500. [CrossRef] [PubMed]
75. O'Sullivan, A.M.; O'Callaghan, Y.C.; O'Connor, T.P.; O'Brien, N.M. Comparison of the Antioxidant Activity of Commercial Honeys, Before and After In-Vitro Digestion. *Polish J. Food Nutr. Sci.* **2013**, *63*, 167–171. [CrossRef]
76. Pan, F.F.; Shao, J.; Shi, C.J.; Li, Z.P.; Fu, W.M.; Zhang, J.F. Apigenin promotes osteogenic differentiation of mesenchymal stem cells and accelerates bone fracture healing via activating Wnt/β-catenin signaling. *Am. J. Physiol. Endocrinol. Metab.* **2021**, *320*, E760–E771. [CrossRef]
77. Huo, J.F.; Zhang, M.L.; Wang, X.X.; Zou, D.H. Chrysin induces osteogenic differentiation of human dental pulp stem cells. *Exp. Cell Res.* **2021**, *400*, 112466. [CrossRef]
78. Aloud, A.A.; Chinnadurai, V.; Govindasamy, C.; Alsaif, M.A.; Al-Numair, K.S. Galangin, a dietary flavonoid, ameliorates hyperglycaemia and lipid abnormalities in rats with streptozotocin-induced hyperglycaemia. *Pharm. Biol.* **2018**, *56*, 302–308. [CrossRef]
79. Dixon, R.A.; Ferreira, D. Genistein. *Phytochemistry* **2002**, *60*, 205–211. [CrossRef]
80. Parim, B.; Sathibabu Uddandrao, V.V.; Saravanan, G. Diabetic cardiomyopathy: Molecular mechanisms, detrimental effects of conventional treatment, and beneficial effects of natural therapy. *Heart Fail. Rev.* **2019**, *24*, 279–299. [CrossRef]
81. Chen, L.Y.; Cheng, H.L.; Kuan, Y.H.; Liang, T.J.; Chao, Y.Y.; Lin, H.C. Therapeutic Potential of Luteolin on Impaired Wound Healing in Streptozotocin-Induced Rats. *Biomedicines* **2021**, *9*, 761. [CrossRef]
82. Kim, H.Y.; Park, S.Y.; Choung, S.Y. Enhancing effects of myricetin on the osteogenic differentiation of human periodontal ligament stem cells via BMP-2/Smad and ERK/JNK/p38 mitogen-activated protein kinase signaling pathway. *Eur. J. Pharmacol.* **2018**, *834*, 84–91. [CrossRef]
83. Kumar, S.; Pandey, A. Chemistry and biological activities of flavonoids: An overview. *Sci. Tech. Rep.* **2013**, *2013*, 162750. [CrossRef] [PubMed]
84. Natsume, N.; Yonezawa, T.; Woo, J.T.; Teruya, T. Effect of pinocembrin isolated from Alpinia zerumbet on osteoblast differentiation. *Cytotechnology* **2020**, *73*, 307–317. [CrossRef] [PubMed]

85. Casado-Díaz, A.; Anter, J.; Dorado, G.; Quesada-Gómez, J.M. Effects of quercetin, a natural phenolic compound, in the differentiation of human mesenchymal stem cells (MSC) into adipocytes and osteoblasts. *J. Nutr. Biochem.* **2016**, *32*, 151–162 [CrossRef] [PubMed]
86. Shi, H.; Dong, L.; Dang, X.; Liu, Y.; Jiang, J.; Wang, Y.; Lu, X.; Guo, X. Effect of chlorogenic acid on LPS-induced proinflammatory signaling in hepatic stellate cells. *Inflamm. Res.* **2013**, *62*, 581–587. [CrossRef] [PubMed]
87. Tsai, K.L.; Hung, C.H.; Chan, S.H.; Hsieh, P.L.; Ou, H.C.; Cheng, Y.H.; Chu, P.M. Chlorogenic Acid Protects Against oxLDL-Induced Oxidative Damage and Mitochondrial Dysfunction by Modulating SIRT1 in Endothelial Cells. *Mol. Nutr. Food Res.* **2018**, *62*, e1700928. [CrossRef]
88. Shimoyama, A.T.; Santin, J.R.; Machado, I.D.; de Oliveira e Silva, A.M.; de Melo, I.L.; Mancini-Filho, J.; Farsky, S.H. Antiulcerogenic activity of chlorogenic acid in different models of gastric ulcer. *Naunyn Schmiedebergs Arch. Pharmacol.* **2013**, *386*, 5–14. [CrossRef]
89. Ornano, L.; Venditti, A.; Donno, Y.; Sanna, C.; Ballero, M.; Bianco, A. Phytochemical analysis of non-volatile fraction of *Artemisia caerulescens* subsp. densiflora (Viv.) (Asteraceae), an endemic species of La Maddalena Archipelago (Sardinia--Italy). *Nat. Prod. Res.* **2016**, *30*, 920–925. [CrossRef]
90. Venditti, A.; Maggi, F.; Vittori, S.; Papa, F.; Serrilli, A.M.; Di Cecco, M.; Ciaschetti, G.; Mandrone, M.; Poli, F.; Bianco, A. Antioxidant and α-glucosidase inhibitory activities of *Achillea tenorii*. *Pharm. Biol.* **2015**, *53*, 1505–1510. [CrossRef]
91. Venditti, A.; Guarcini, L.; Bianco, A.; Rosselli, S.; Bruno, M.; Senatore, F. Phytochemical analysis of Achillea ligustica All. from Lipari Island (Aeolian Islands). *Nat. Prod. Res.* **2016**, *30*, 912–919. [CrossRef]
92. Sharifi-Rad, M.; Nazaruk, J.; Polito, L.; Morais-Braga, M.F.B.; Rocha, J.E.; Coutinho, H.D.M.; Salehi, B.; Tabanelli, G.; Montanari, C.; Del Mar Contreras, M.; et al. Matricaria genus as a source of antimicrobial agents: From farm to pharmacy and food applications. *Microbiol. Res.* **2018**, *215*, 76–88. [CrossRef]
93. Venditti, A.; Frezza, C.; Sciubba, F.; Serafini, M.; Bianco, A.; Cianfaglione, K.; Maggi, F. Volatile components, polar constituents and biological activity of tansy daisy (*Tanacetum macrophyllum* (Waldst. et Kit.) Schultz Bip. *Ind. Crops Prod.* **2018**, *118*, 225–235. [CrossRef]
94. Zhang, X.; Zhou, C.; Zha, X.; Xu, Z.; Li, L.; Liu, Y.; Xu, L.; Cui, L.; Xu, D.; Zhu, B. Apigenin promotes osteogenic differentiation of human mesenchymal stem cells through JNK and p38 MAPK pathways. *Mol. Cell Biochem.* **2015**, *407*, 41–50. [CrossRef] [PubMed]
95. Jaganathan, S.K.; Mandal, M. Antiproliferative effects of honey and of its polyphenols: A review. *J. Biomed. Biotechnol.* **2009**, *2009*, 830616. [CrossRef] [PubMed]
96. Anand, K.V.; Anandhi, R.; Pakkiyaraj, M.; Geraldine, P. Protective effect of chrysin on carbon tetrachloride (CCl4)-induced tissue injury in male Wistar rats. *Toxicol. Ind. Health* **2011**, *27*, 923–933. [CrossRef] [PubMed]
97. Anand, K.V.; Mohamed Jaabir, M.S.; Thomas, P.A.; Geraldine, P. Protective role of chrysin against oxidative stress in d-galactose-induced aging in an experimental rat model. *Geriatr. Gerontol. Int.* **2012**, *12*, 741–750. [CrossRef]
98. Amin Altawash, A.S.; Shahneh, A.Z.; Moravej, H.; Ansari, M. Chrysin-induced sperm parameters and fatty acid profile changes improve reproductive performance of roosters. *Theriogenology* **2017**, *104*, 72–79. [CrossRef]
99. Hadjmohammadi, M.R.; Nazari, S.S. Separation optimization of quercetin, hesperetin and chrysin in honey by micellar liquid chromatography and experimental design. *J. Sep. Sci.* **2010**, *33*, 3144–3151. [CrossRef]
100. Zhao, B.; Xiong, Y.; Zhang, Y.; Jia, L.; Zhang, W.; Xu, X. Rutin promotes osteogenic differentiation of periodontal ligament stem cells through the GPR30-mediated PI3K/AKT/mTOR signaling pathway. *Exp. Biol. Med.* **2020**, *245*, 552–561. [CrossRef]
101. Habryka, C.; Socha, R.; Juszczak, L. The Effect of Enriching Honey with Propolis on the Antioxidant Activity, Sensory Characteristics, and Quality Parameters. *Molecules* **2020**, *25*, 1176. [CrossRef]
102. Martos, I.; Ferreres, F.; Yao, L.; D'Arcy, B.; Caffin, N.; Tomás-Barberán, F.A. Flavonoids in monospecific eucalyptus honeys from Australia. *J. Agric. Food Chem.* **2000**, *48*, 4744–4748. [CrossRef]
103. Elbatreek, M.H.; Mahdi, I.; Ouchari, W.; Mahmoud, M.F.; Sobeh, M. Current advances on the therapeutic potential of pinocembrin: An updated review. *Biomed. Pharmacother.* **2023**, *157*, 114032. [CrossRef] [PubMed]
104. Jasicka-Misiak, I.; Gruyaert, S.; Poliwoda, A.; Kafarski, P. Chemical profiling of polyfloral Belgian honey: Ellagic acid and pinocembrin as antioxidants and chemical markers. *J. Chem.* **2017**, *2017*, 1–8. [CrossRef]
105. Alvear, M.; Santos, E.; Cabezas, F.; Pérez-SanMartín, A.; Lespinasse, M.; Veloz, J. Geographic Area of Collection Determines the Chemical Composition and Antimicrobial Potential of Three Extracts of Chilean Propolis. *Plants* **2021**, *10*, 1543. [CrossRef] [PubMed]
106. Yayinie, M.; Atlabachew, M.; Tesfaye, A.; Hilluf, W.; Reta, C.; Alemneh, T. Polyphenols, flavonoids, and antioxidant content of honey coupled with chemometric method: Geographical origin classification from Amhara region, Ethiopia. *Int. J. Food Prop.* **2022**, *25*, 76–92. [CrossRef]
107. Zhou, Y.; Wu, Y.; Jiang, X.; Zhang, X.; Xia, L.; Lin, K.; Xu, Y. The Effect of Quercetin on the Osteogenesic Differentiation and Angiogenic Factor Expression of Bone Marrow-Derived Mesenchymal Stem Cells. *PLoS ONE* **2015**, *10*, e0129605. [CrossRef]
108. Ardalani, H.; Vidkjær, N.H.; Laursen, B.B.; Kryger, P.; Fomsgaard, I.S. Dietary quercetin impacts the concentration of pesticides in honey bees. *Chemosphere* **2021**, *262*, 127848. [CrossRef]
109. Kędzierska-Matysek, M.; Stryjecka, M.; Teter, A.; Skałecki, P.; Domaradzki, P.; Florek, M. Relationships between the Content of Phenolic Compounds and the Antioxidant Activity of Polish Honey Varieties as a Tool for Botanical Discrimination. *Molecules* **2021**, *26*, 1810. [CrossRef]

110. Lu, H.; Tian, Z.; Cui, Y.; Liu, Z.; Ma, X. Chlorogenic acid: A comprehensive review of the dietary sources, processing effects, bioavailability, beneficial properties, mechanisms of action, and future directions. *Compr. Rev. Food Sci. Food Saf.* **2020**, *19*, 3130–3158. [CrossRef]
111. Heitman, E.; Ingram, D.K. Cognitive and neuroprotective effects of chlorogenic acid. *Nutr. Neurosci.* **2017**, *20*, 32–39. [CrossRef]
112. Wang, J.; Xue, X.; Du, X.; Cheng, N.; Chen, L.; Zhao, J.; Zheng, J.; Cao, W. Identification of acacia honey adulteration with rape honey using liquid chromatography-electrochemical detection and chemometrics. *Food Anal. Methods* **2014**, *7*, 2003–2012. [CrossRef]
113. Bonsignore, G.; Martinotti, S.; Ranzato, E. Wound Repair and Ca. *Cells* **2024**, *13*, 491. [CrossRef] [PubMed]
114. Erejuwa, O.O. Effect of honey in diabetes mellitus: Matters arising. *J. Diabetes Metab. Disord.* **2014**, *13*, 23. [CrossRef] [PubMed]
115. Afrin, S.; Haneefa, S.M.; Fernandez-Cabezudo, M.J.; Giampieri, F.; Al-Ramadi, B.K.; Battino, M. Therapeutic and preventive properties of honey and its bioactive compounds in cancer: An evidence-based review. *Nutr. Res. Rev.* **2020**, *33*, 50–76. [CrossRef]
116. Bonsignore, G.; Patrone, M.; Grosso, F.; Martinotti, S.; Ranzato, E. Cancer Therapy Challenge: It Is Time to Look in the "St. Patrick's Well" of the Nature. *Int. J. Mol. Sci.* **2021**, *22*, 10380. [CrossRef] [PubMed]
117. Martinotti, S.; Pellavio, G.; Patrone, M.; Laforenza, U.; Ranzato, E. Manuka Honey Induces Apoptosis of Epithelial Cancer Cells through Aquaporin-3 and Calcium Signaling. *Life* **2020**, *10*, 256. [CrossRef]
118. Martinotti, S.; Bonsignore, G.; Ranzato, E. Understanding the Anticancer Properties of Honey. *Int. J. Mol. Sci.* **2024**, *25*, 11724. [CrossRef]
119. Khalil, M.I.; Sulaiman, S.A. The potential role of honey and its polyphenols in preventing heart diseases: A review. *Afr. J. Tradit. Complement. Altern. Med.* **2010**, *7*, 315–321. [CrossRef]
120. Bt Hj Idrus, R.; Sainik, N.Q.A.V.; Nordin, A.; Saim, A.B.; Sulaiman, N. Cardioprotective Effects of Honey and Its Constituent: An Evidence-Based Review of Laboratory Studies and Clinical Trials. *Int. J. Environ. Res. Public. Health* **2020**, *17*, 3613. [CrossRef]
121. Obi, C.L.; Ugoji, E.O.; Edun, S.A.; Lawal, S.F.; Anyiwo, C.E. The antibacterial effect of honey on diarrhoea causing bacterial agents isolated in Lagos, Nigeria. *Afr. J. Med. Med. Sci.* **1994**, *23*, 257–260.
122. al Somal, N.; Coley, K.E.; Molan, P.C.; Hancock, B.M. Susceptibility of Helicobacter pylori to the antibacterial activity of manuka honey. *J. R. Soc. Med.* **1994**, *87*, 9–12. [CrossRef]
123. Ndip, R.N.; Malange Takang, A.E.; Echakachi, C.M.; Malongue, A.; Akoachere, J.F.; Ndip, L.M.; Luma, H.N. In-vitro antimicrobial activity of selected honeys on clinical isolates of Helicobacter pylori. *Afr. Health Sci.* **2007**, *7*, 228–232. [PubMed]
124. Bonsignore, G.; Patrone, M.; Martinotti, S.; Ranzato, E. "Green" Biomaterials: The Promising Role of Honey. *J. Funct. Biomater.* **2021**, *12*, 72. [CrossRef] [PubMed]
125. Puścion-Jakubik, A.; Borawska, M.H.; Socha, K. Modern Methods for Assessing the Quality of Bee Honey and Botanical Origin Identification. *Foods* **2020**, *9*, 1028. [CrossRef] [PubMed]
126. Tremblay, P.; Paquin, R. Improved detection of sugar addition to maple syrup using malic acid as internal standard and in 13C isotope ratio mass spectrometry (IRMS). *J. Agric. Food Chem.* **2007**, *55*, 197–203. [CrossRef]
127. Çinar, S.B.; Ekşi, A.; Coşkun, İ. Carbon isotope ratio (13C/12C) of pine honey and detection of HFCS adulteration. *Food Chem.* **2014**, *157*, 10–13. [CrossRef]
128. Majtan, J. testing of honey quality and biological functionality: Underestimated elements in the clinical testing of honey. *Front. Nutr.* **2024**, *11*, 1433786. [CrossRef]

BioTech

MDPI

Review

Prospects in the Use of *Cannabis sativa* Extracts in Nanoemulsions

Ian Vitola [1], Carlos Angulo [2,*], Raul C. Baptista-Rosas [3,4], Luis Miguel Anaya-Esparza [5], Zazil Yadel Escalante-García [1], Angélica Villarruel-López [6] and Jorge Manuel Silva-Jara [6,*]

1 Departamento de Ingeniería Química, Universidad de Guadalajara, CUCEI, Blvd. Marcelino García Barragán 1421, Olímpica, Guadalajara 44430, Jalisco, Mexico; ian.castro8806@alumnos.udg.mx (I.V.); zazil.esclante@academicos.udg.mx (Z.Y.E.-G.)
2 Grupo de Inmunología y Vacunología, Centro de Investigaciones Biológicas del Noroeste, S.C. (CIBNOR), Instituto Politécnico Nacional 195, Playa Palo de Santa Rita Sur, La Paz 23096, Baja California Sur, Mexico
3 Departamento de Ciencias de la Salud-Enfermedad como Proceso Individual, CUTonalá, Universidad de Guadalajara, Nuevo Perif. Ote. 555, Ejido San José, Tateposco, Tonalá 45425, Jalisco, Mexico; raul.baptista@academicos.udg.mx
4 Hospital General de Occidente, Secretaría de Salud Jalisco, Av. Zoquipan 1050, Colonia Zoquipan, Zapopan 45170, Jalisco, Mexico
5 Centro de Estudios Para la Agricultura, la Alimentación y la Crisis Climática, Centro Universitario de los Altos, Universidad de Guadalajara, Rafael Casillas Aceves 1200, Tepatitlán de Morelos 47600, Jalisco, Mexico; luis.aesparza@academicos.udg.mx
6 Departamento de Farmacobiología, Universidad de Guadalajara, CUCEI, Blvd. Marcelino García Barragán 1421, Olímpica, Guadalajara 44430, Jalisco, Mexico; angelica.vlopez@academicos.udg.mx
* Correspondence: eangulo@cibnor.mx (C.A.); jorge.silva@academicos.udg.mx (J.M.S.-J.)

Abstract: *Cannabis sativa* plants have been widely investigated for their specific compounds with medicinal properties. These bioactive compounds exert preventive and curative effects on non-communicable and infectious diseases. However, *C. sativa* extracts have barely been investigated, although they constitute an affordable option to treat human diseases. Nonetheless, antioxidant, antimicrobial, and immunogenicity effects have been associated with *C. sativa* extracts. Furthermore, innovative extraction methods in combination with nanoformulations have been proposed to increase desirable compounds' availability, distribution, and conservation, which can be aided by modern computational tools in a transdisciplinary approach. This review aims to describe available extraction and nanoformulation methods for *C. sativa,* as well as its known antioxidant, antimicrobial, and immunogenic activities. Critical points on the use of *C. sativa* extracts in nanoformulations are identified and some prospects are envisaged.

Keywords: medicinal plants; in silico analysis; low-cost medicine; alternative uses; cannabidiol

Key Contribution: This review article provides a comprehensive view of *Cannabis sativa*'s potential in modern medicine, highlighting its antioxidant, antimicrobial, and immunostimulant properties; computational tools; and innovations in its extraction and formulation at the nanoscale level.

1. Introduction

Cannabis is a plant that belongs to the *Cannabaceae* family; it is annual and dioecious. Over time, since its appearance (possibly more than 5000 years ago), it has been used for purposes such as obtaining fibers and oils. Originally, *Cannabis sativa* is believed to be indigenous to Eastern Asia, particularly in Mongolia and southern Siberia [1]. Due to its cultivation for various uses, this plant has spread globally and is now found in many parts of the world, including America, Europe, and Asia. Its adaptability to different climates and soils has contributed to its cosmopolitan distribution. The plant thrives in diverse environments, which has led to significant morphological and chemical variations based on geographical location and cultivation practices. At the pharmacological level, compounds such as alkaloids, flavonoids, terpenoids, and cannabinoids stand out. Of the latter, more than 130 have been isolated in trichomes of *Cannabis sativa* [2]. The structures in trichomes can be divided into glandular and non-glandular. Within the non-glandular trichomes, two types are not linked to the production of terpenoids, whereas there are three types of trichomes in the glandular ones: (1) stalked, (2) sessile, and (3) bulbous [3,4]. They are made up of secretory cells and stems with different functions. In recent years, there has been a resurgence of interest in its therapeutic potential due to its complex chemical composition, including cannabinoids such as tetrahydrocannabinol (THC) and cannabidiol (CBD), among others [5]. Although several authors have proposed classifying them into three species, *C. sativa*, *Cannabis indica*, and *Cannabis ruderalis*, most research groups agree that it is a monotypic plant; that is, all varieties belong to the same species, *C. sativa*. Another classification is the chemotaxonomic one, which considers phenotypes based on the proportion of cannabinoids. However, even though this classification may not fully account for the cannabinoid content that different chemotypes can present, it is still considered a helpful tool for classification [6].

The *Cannabis sativa* plant, a subject of extensive research, holds immense promise in the modulation of the immune system [7], oxidative stress [8], and antimicrobial capacity [9]. These studies have revealed the potential of cannabinoids, terpenes, flavonoids, and lignans [10], which are metabolites present in the plant that can bolster the body's antioxidant response, enhance antimicrobial treatments, and exert beneficial immunomodulatory effects [11–13]. Unlike isolated compounds, these phytochemicals present in extracts can also act synergistically to promote therapeutic effects, a phenomenon known as the entourage effect [14]. With approximately 500 compounds in *C. sativa*, the presence of THC in extracts has been a critical focus, leading to the study of cannabinoids such as CBD [15] and other terpenes such as β-caryophyllene (BCP). The potential benefits of these extracts are vast, and overcoming the challenges in their practical application, such as reducing the psychoactive effect of THC [16], preventing the degradation of these active compounds [17,18], and enhancing bioavailability [19], could revolutionize the field of *C. sativa* therapeutics.

The problem of the psychotropic stimulus caused by THC has been addressed by exploring other non-psychoactive cannabinoids, such as CBD, cannabigerol (CBG), cannabichromene (CBC), and cannabidivarin (CBDV) [20]. On the other hand, to reduce the degradation processes of active compounds, the development of protection methods such as nanoencapsulation has been promoted [21]. This approach protects cannabinoids, terpenes, and flavonoids from environmental effects (light, temperature, and oxygen) to prolong their stability and half-life [22]. Related to the above, the improvement of bioavailability has been studied to increase the therapeutic capacity of active compounds, highlighting that the oral route has the lowest assimilation rate [23,24]. For example, nanoemulsions are an alternative to the challenge of oral administration because they can protect bioactive compounds from digestive processes and improve their bioavailability and bioaccessibility.

Nanoemulsions have been manufactured from oils/lipids (olive oil, phospholipids, and sunflower oil) [25,26], proteins (albumins, soy protein isolate, whey protein isolate, β-lactoglobulin, rice protein isolate, and peanut protein) [27–29], polysaccharides (carboxymethylcellulose, chitosan, glucans, pectins, alginates, hyaluronic acid, starches, cellulose, xanthan, guar, and gum arabic) [30–34], surfactants (sodium lecithin, lactoglobulins, and sodium dodecyl sulfate) [35], and plasticizers (glycerol, sorbitol, propylene glycol, and polyethylene glycol) [36,37], which form nanostructures with improved solubility and absorption of bioactive compounds. This technology can protect bioactive compounds from *C. sativa* for diverse therapeutic purposes. In this context, this paper describes and discusses recent research on *C. sativa* extracts (ECs) and the potential of nanoemulsion formulation to protect bioactive compounds and improve their bioavailability and bioaccessibility.

2. Extraction Methods Reported for *C. sativa*

C. sativa is considered an excellent source of bioactive compounds (cannabinoids, terpenes, flavonoids, and lignans) that can be extracted through diverse techniques [38].

Extraction methods can be classified into three broad groups. Traditional methods are based on the use of solvents (such as Soxhlet and maceration), alternative methods [38], and conventional methods without the use of solvents [39]. The main extraction methods used for *Cannabis sativa* are depicted in Figure 1. Solvent-based extractions can be subclassified into those using alcohols (methanol and ethanol), hexane, and ethers [40,41]. In this type of extraction, the solid material crushed in the organic solvent is disposed of in different proportions, filtered, and stored. In extractions using ether, organic acids are used to adjust the pH [42], and continuous evaporation–condensation cycles can be used to improve efficiency [43] or dynamic maceration, which is more efficient [44]. In addition, there are other methods for making oil extractions in which the use of edible vegetable oils as a solvent is required, such as coconut [45], sunflower [46], and olive [47]. For example, *Cannabis* flowers are dried between 85 and 145 °C for 40 min in a convection oven, pulverized, coated with olive oil (1:10), stirred for 40 to 120 min, filtered, and stored [47]. Although widely used, this type of extraction is inefficient and, in some cases, can leave potentially toxic residues [48].

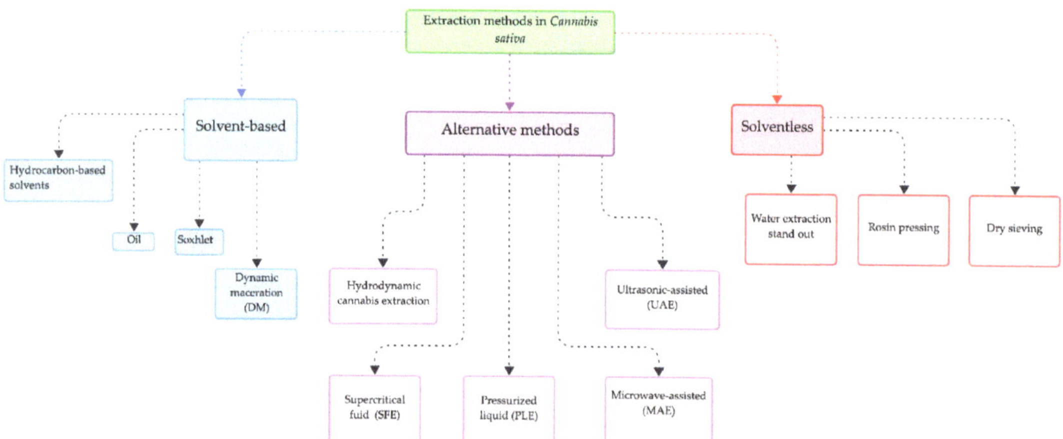

Figure 1. Extraction methods for *Cannabis*.

Alternative extraction methods differ from traditional ones mainly because of their higher efficiency. They use physical and chemical principles that optimize the process, allowing its use in the food, cosmetic, and research industries. There are ultrasound-assisted (UAE), microwave-assisted (MAE), pressurized liquid (PLE), supercritical fluid (SFE) [49], and *Cannabis* hydrodynamics extractions [50]. In the case of PLE and SFE extractions, CO_2 and high pressure are used [51] to achieve better extractions [52]. For example, the conditions that have been recommended for cannabinoid extraction are $T \geq 31$ °C, $P \geq 1015$ psi for CO_2, $T \geq 57$ °C, $P \geq 2625$ psi for CO_2/Absolute Ethanol (95/5, v/v) at a flow rate of 3 mL/min and (a) an extraction time of 60 min for SFE and 4 mL/min and (b) 30 min of extraction for PLE [53]. On the other hand, ultrasound-assisted (UAE) and microwave-assisted (MAE) extraction techniques are similar to the previous ones in that they reduce the time and use of solvents while obtaining bioactive compounds [54]. UAE extraction uses acoustic cavitation to generate microbubbles that break up plant material, and MAE extraction uses rapid microwave heating to increase the solubility of compounds and facilitate the extraction of cannabinoids, terpenes, and other compounds of interest using sample–solvent ratios, times, and temperatures [55]. Finally, the hydrodynamic extraction of *Cannabis* requires freezing the plant material, converting it into a nanoemulsion using hot water, and using ultrasound to use the hydrodynamic force to break the cell wall; finally, the steps they follow are liquid–liquid extraction, centrifugation, and drying at low temperatures [56]. Despite having the name "Cannabis", this method has been little used on this plant. For a generalized visual summary of extraction methods for *Cannabis sativa*, see Figure 2.

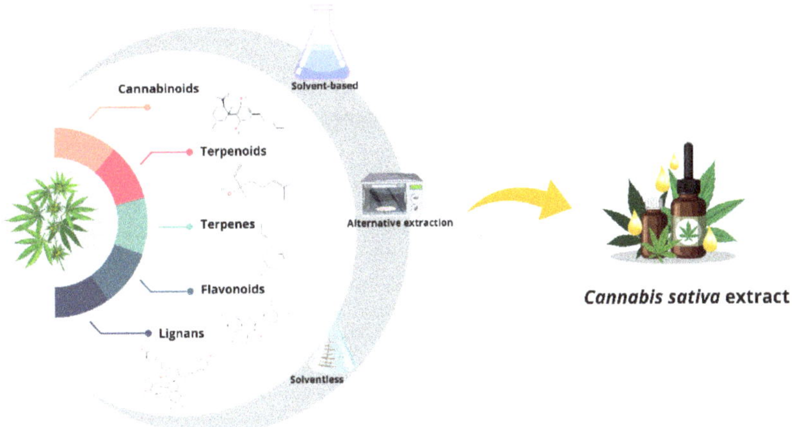

Figure 2. Methods that are traditional, alternative, and solvent-free. Each process can influence the efficiency and purity of the *Cannabis*'s bioactive compounds.

Some critical extraction points may be considered in the extraction methods for obtaining bioactive compounds or extracts from *C. sativa*, as discussed below.

The solvent is decisive in the extraction of bioactive compounds; the efficiency and quality of the product depend on it, and the compound to be used must be safe and environmentally friendly. Solvents such as hexane, butane, and ether have been widely used. However, they are highly flammable and toxic, representing potential health risks [48]. These solvents can leave traces in extracts, and synthesizing nanoemulsions with them can affect purity, safety, and quality [22]. Therefore, choosing a solvent that achieves a high extraction performance, is safe, and is environmentally friendly is a challenge. Supercritical solvents have been a convenient alternative; however, high costs can reduce their use widely [57]. On the other hand, vegetable oils and solvents offer the possibility of obtaining less expensive quality extracts that do not compromise human health or pose a danger to the environment. In addition, simultaneous use with other technologies, such as ultrasonic extraction, could improve performance [58].

The extraction process requires controlling variables like extraction time, temperature, pressure, the proportion of plant material and solvent, and stirring speed [59]. Carelessness in one of these variables could lead to the loss of sensitive compounds. In this sense, reproducing an optimal and quality process is difficult, representing a drawback. Automation strategies to optimize and control the critical variables of the extraction process could be a good target for exploration, thus constantly adjusting the conditions as required by the process with precision and efficiency and avoiding less loss of bioactive compounds, which translates into better extraction percentages and the quality of the phytochemical profiles of interest [60,61].

3. Nanoemulsions: Materials and Synthesis Methods

Nanoemulsions consist of the dispersion of two immiscible liquid phases, where one is dispersed in the other in the form of droplets on a nanometric scale (<100 nm). They are composed of an aqueous phase, an oil phase, and an emulsifier. Nanoemulsions are primarily prepared with an aqueous phase, an oily phase, and surfactants [62]. Polar materials frequently used in the aqueous phase include polyols [63], simple alcohols [64], proteins [65], and carbohydrates [66]. In the oil phase, acylglycerols [67], essential oils [68], vitamins [69], and others of a lipophilic nature [70] are used. Surfactants confer stability to nanoemulsions and include lecithin [71], lactoglobulin [72], sodium dodecyl sulfate [73], and saponin [37,74]. Although these materials have been shown to protect cannabinoids and other compounds of interest from *C. sativa*, the formulation must be optimized, and the release mechanisms must be considered.

Additionally, surfactants, or emulsifiers, play an essential role in stabilizing nanoemulsions. They hold the two immiscible phases, oily and aqueous, together. However, the fillers of bioactive compounds can interfere with the emulsifier or have a greater affinity for one of the phases, causing processes such as Ostwald maturation in the short term [54]. Concerning synthesis, each stage of the process deserves care, starting from the component's homogenization to the fine emulsion's final packaging. Factors such as temperature, mixing order, stirring time, and speed in the homogenization process influence droplet size and polydispersity [36]. In this way, a good experimental design is necessary, and the quality of the equipment must be guaranteed throughout the process.

The synthesis of nanoemulsions is not the last step of the process; stability is one of the most critical points, as the particle size, the zeta potential, the surfactant, and the composition of the dispersing phase condition it. Cremation, sedimentation, coalescence, and Ostwald maturation are consequences of a loss of stability and bioactive compounds [75]. If large nanoemulsion droplets lead to sedimentation/cremation, an unfavorable zeta potential leads to aggregation and a marked difference in droplet sizes. Aspects such as viscosity can affect how the emulsion can be applied or dosed in experimental animal models [54]. To state the above in physicochemical terms, at a biological level, nanoemulsions could present little assimilation if the components used are not related to the cells of the target tissue. The stability and viscosity of nanoemulsions should be explored with new surfactants that reduce the interfacial tensions of the phases without significantly increasing the nanoemulsion's size or the continuous phase's rheological properties. Sedimentation, on the other hand, the surface modification of the particles, could improve electrostatic repulsion and prevent the aggregation of droplets. However, it must be ensured that this functionalization does not alter the mechanism of action of the nanoemulsion [76]. Furthermore, the synthesis method significantly influenced nanoemulsion properties. In this context, high- and low-energy methods have been used to synthesize nanoemulsions (Figure 3), as described below.

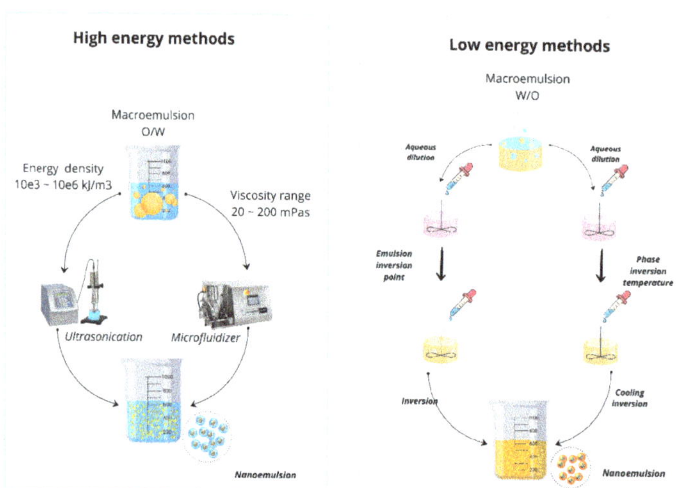

Figure 3. Nanoemulsion synthesis methods. High- and low-energy methods are the two groups into which they are classified. High-energy methods: the cavitation produced by the high frequency of the

ultrasonic equipment breaks the emulsion droplets into smaller ones. Microfluidizers use high pressures that break up emulsion droplets. Low-energy methods: the emulsion inversion point (EIP) consists of changing the nature of the emulsion from water in oil (W/O) to oil in water (O/W), modifying the concentration and type of surfactants, varying the speed and inversion temperatures [54,77].

3.1. High-Energy Methods Used for Nanoemulsions

High extraction methods require high-intensity forces and little use of surfactants to cause an increase in kinetic energy that breaks droplets in the oil phase and increases dispersion in the aqueous phase [78]. Commonly used high-energy methods are ultrasound, high-pressure homogenization, and microfluidization. Ultrasound uses high-frequency waves in the mixture, generating cavitation and the collapse of the microparticles, reducing the size to nanometer values. It is necessary to adjust the pulse's time and intensity since the temperature increase can cause the degradation of the compounds of interest [54,79]. High-pressure homogenization passes the mixture through a small hole that breaks the droplets to nanometer-scale value sizes, so a precise flow rate and pressure control are required [54,80]. Finally, microfluidization makes the mixture transit at high speed through small channels to cause shear forces capable of breaking the droplets of the emulsions, turning them into nanoemulsions. Characteristics such as channel geometry and velocity determine conditions for achieving normal size distributions in nanoemulsions [54,81].

3.2. Low-Energy Methods Used for Nanoemulsions

Low-energy methods for nanoemulsions are recommended to avoid the degradation of thermolabile compounds. However, they may need other types of surfactants and a greater quantity to stabilize the nanoemulsions [54,78]. The most used are solvent evaporation, membrane emulsification, and spontaneous emulsification. In solvent evaporation, the organic solvent in the mixture, in which the oil phase and surfactants are located, evaporates in a controlled manner. Selecting the appropriate solvent is critical, and one with a low boiling point is recommended because it allows the use of low temperatures, such as ethanol, acetone, lactic-co-glycolic acid (PGLA) [82], and tetrahydrofuran [70]. Another critical consideration to review is solvent safety and compatibility. In membrane emulsification, the dispersion of the oil phase is achieved by forcing the mixture through a porous membrane with electrical or hydrodynamic forces. The main element of this method is the selection of the membrane with appropriate pore sizes, pressure values, flow rate, and temperature control to obtain stable and uniform nanoemulsions [83]. Spontaneous emulsification is achieved by correctly combining the components of the mixture so that, under specific conditions of pH, temperature, and concentration, stable and uniform nanoemulsions are formed without needing energy [84,85].

In general, the nanoemulsions achieved by these methods are stable and uniform, which is suitable for applications in the pharmaceutical industry. In addition, the combination of different techniques can be considered, which can improve stability and biodistribution. The possibility of preserving bioactive compounds and their properties drives continuous process refinement and optimization based on the size, distribution, stability, and controlled release of *C. sativa* emulsions.

4. Antioxidant Activity of *Cannabis sativa* Extracts and Nanoemulsions

The study of cannabinoids has become an area of great interest due to their therapeutic potential, in which antioxidant power stands out. Although *C. sativa* extracts (CSEs) contain several cannabinoids, little is known about their antioxidant properties acting together. No reports were found on the analysis of the antioxidant capacity of CSE, although some bioactive compounds reported in the plant were found. In this context, it must be noted that the antioxidant capacity of extracts is essential for therapeutic applications, but determining that capacity can be challenging due to the heterogeneity of bioactive compounds [86].

The antioxidant activity of CBD increases the concentration of anandamide that activates the peroxisome proliferator alpha (PPAR-α), which in turn participates in the regulation of the expression of the antioxidant enzyme superoxide dismutase [87]. In another study, CBD decreased reactive oxygen species (ROS), pro-inflammatory cytokines, and lipid peroxidation in LPS-stimulated C57BL/6J mouse microglial cells [88]. Similarly, CSE with 72% THC reduced ROS formation by 80% in differentiated SH-SY5Y neurons compared to purified CBD and THC [89]. Not only are cannabinoids found in CSE that may have antioxidant properties, but flavonoids, terpenes, and other phytochemicals known for their antioxidant properties are also present [90]. Because the bioavailability of cannabinoids varies according to the route of administration (i.e., 6% orally) [91], nanoemulsions could improve the stability and bioavailability of phytocannabinoids and other CSE compounds with antioxidant capacity; other studies have evaluated nanoemulsions of crude extracts or compounds with antioxidant capacity. In this regard, it was demonstrated that a nanoemulsion with a *Nigella sativa* seed extract containing thymoquinone had greater bioavailability and antioxidant capacity than pure extract. In the context of *C. sativa* crude extract, this has been evaluated for its antioxidant properties in nanoemulsions, the opposite of cannabinoids such as CBD, which has been widely studied [92–96]. Therefore, studying nanoemulsions with CSE to increase antioxidant capacity is an opportunity for future research; all the above is represented in Figure 4.

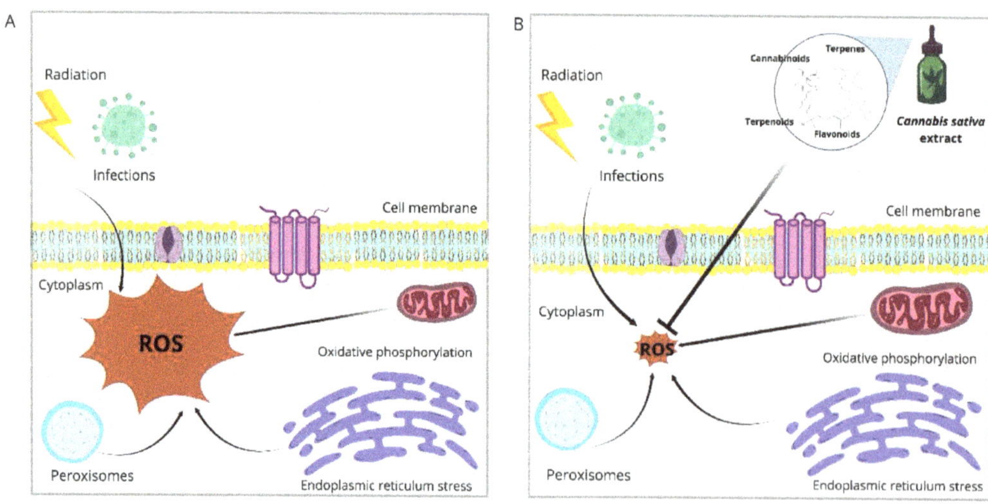

Figure 4. Antioxidant properties. ROS: reactive oxygen species. (**A**) Cell with a large amount of ROS produced by different factors. (**B**) The administration of EC is rich in antioxidant compounds capable of reducing reactive oxygen species [97].

5. Effect of *Cannabis sativa* Extract and Nanosystems on Antimicrobial Activity

Among the therapeutic interests aroused by *C. sativa* are its antimicrobial properties. The improved distribution of nanoemulsions shows promise in encapsulated extracts in dose reduction for effective microbial activity, although few studies have been published. For example, extracts of different varieties of *C. sativa* had variable antimicrobial effects depending on their chemical composition and the concentration used on the microbial strains. The most significant inhibition of growth by *C. sativa* extracts was against *Bacillus thuringiensis, Staphylococcus aureus, Candida albicans, Micrococcus luteus, Pseudomosas protegens, Saccharomyces cerevisiae,* and *Fusarium eumartii* [92]. In another study, *C. sativa* seed extract inhibited the growth of pathogenic *Enterobacteriaceae* and the formation of *Staphylococcus aureus* biofilms without detecting an antimicrobial effect on probiotic bacteria of the genus *Bifidobacterium* and *Lactobacillus* [98].

Cannabis extract has been shown to have an antimicrobial effect on pathogenic strains, an ability that may be linked to cannabinoids and terpenes; however, the level of activity can vary according to various factors such as the concentration of the bioactive compounds, the sensitivity of the target strains [99], and the level of nanoemulsions, the ability of the structures to cross the membranes of the microorganisms, the rate of erosion of the nanoemulsion, and the action against the efflux pumps present in the bacteria [100]. Thus, it is necessary to determine the antimicrobial capacity in several contexts to know the antimicrobial activity level of the nanoemulsion. In this context, it is crucial to determine the antimicrobial capacity and the spectrum of action of nanoemulsion-based *C. sativa* extracts. Bioactive compounds can be specific to Gram-negative or Gram-positive bacteria, non-selective, and harmful to beneficial bacteria [98]. However, studies on specificity, concentration variation in nanoemulsions, and spectrum range are suggested by testing on various pathogenic and beneficial Gram-positive and Gram-negative strains. In this way, the mechanism of antimicrobial action can be better understood, allowing better applications in treating pathogenic strains or promoting beneficial ones.

Additionally, an innovative approach has been the biosynthesis of *C. sativa* nanoparticles with antimicrobial activities. Fungicidal effects were studied using solid lipid nanoparticles (SLNs) and chitosan against the fungus *Fusarium solani* with a lower concentration of pure extract [101]. The pure EC obtained consisted of 770.3 mg of THC, 72.1 mg of CBD, and 28 mg of CBN per gram of resin and various terpenes (α-pinene 9338 mg, β-pinene 3132 mg, β-myrcene 29,837 mg, p-cymene 4031 mg, ocimene 2.3 mg, linalool 2853 mg, BCP 7038 mg, and α-humulene 1660 mg per gram of resin) and chitosan SLNs interacted with the negative charges of cell membranes, facilitating release into cells. In this way, the combination of terpenes, cannabinoids, and chitosan exerted a fungicidal effect against the microorganism. The researchers synthesized four types of SLN: control (183 nm), chitosan (237 nm, SLNQ), *Cannabis* (146.9 nm, SLNC), and Cannabis–chitosan (152.4 nm, SLNCQ), with the latter having greater inhibition capacity (90%) at an extract concentration of 0.06 μg/mL. SLNCQs had an EC release rate of 42.4% over 24 h, higher than that found in SLNCs, with 26.4%. In another report, bimetallic gold and silver nanoparticles synthesized with *Vitis vinifera* canes and *C. sativa* residues had antimicrobial activity by reducing the growth and biofilm formation of *Pseudomonas aeruginosa* [102]. Although nanostructures are not strictly speaking nanoemulsions, they are nanosystems and allow them to be considered in exploring other antimicrobial applications. Figure 5 graphically represents all the above.

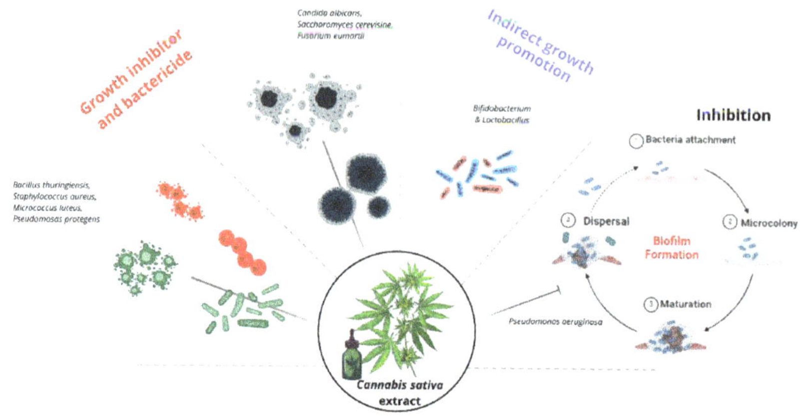

Figure 5. Antimicrobial activity. The EC inhibits bacterial growth, disrupting the bacterial membrane. It can also inhibit spore proliferation, but the effects at a structural level are unknown [98,100–102].

6. Effect of *Cannabis sativa* Extracts Nanoemulsions on Immunogenicity Activity

Another aspect of the therapeutic properties of *C. sativa* to consider is immunomodulation [103–105]; see Figure 6. One of the aspects to highlight is that pure extracts of *C. sativa* have possibly been little studied due to the presence of THC and its known psychotropic effect [106]. There are relatively few studies that include cannabinoid mixtures. Blanton and collaborators [107] reported that the CBD–BCP interaction modulated pain and inflammation in a murine model. In another study, indomethacin (IND) nanoemulsions were prepared with BCP (BCP–IND) in different proportions to test the anti-inflammatory effect in LPS-stimulated macrophages. The results of the trials showed that BCP–IND nanoemulsions reduced the production of pro-inflammatory cytokines more than only BCP or IND. This indicates that cannabinoids alone have therapeutic potential. Still, in conjunction with other cannabinoids or other drugs, they can be even more effective, suggesting the use of pure *C. sativa* extracts in O/W nanoemulsions. In research focused on the treatment of diabetes mellitus (DM), a condition characterized by high oxidative stress and alteration of the immune system [88], Wistar rats with *C. sativa* oil-induced DM (CSO) and nanoemulsions of *C. sativa* oil (NECSO) and metformin were used. The trials demonstrated that CSO and NECSO positively affected DM symptoms and reduced glucose in urine, blood, triglycerides, and low-density lipoproteins (LDL) [108]. A graphic summary of this section is presented in Figure 6.

Although research on *Cannabis* and its great potential to modulate the immune system has been growing, it still faces the difficulty of the wide diversity of bioactive compounds in the plant and its multiple interactions with cells [103,105,110]. Thus, the synergistic effect of around 500 compounds is not the same when isolated. Therefore, characterizing each of these compounds and understanding their therapeutic potential is challenging. Regarding immunomodulation, whether anti- or pro-inflammatory, one bioactive compound could have a favorable effect, while at the same time, another can act as an inverse agonist [111], so elucidating each compound's effects requires rigorous and exhaustive approaches. Preclinical studies addressing in silico, in vitro, and in vivo assays using animal models and clinical studies in humans are a possibility to validate the immunomodulatory effects of extracts. The immune system's high specialization makes it highly complex [112], so it is necessary to evaluate the interactions of these compounds with the different cell lineages and sublineages that make up the system, as well as the nanoemulsions based on *C. sativa* extracts' immunomodulatory effects.

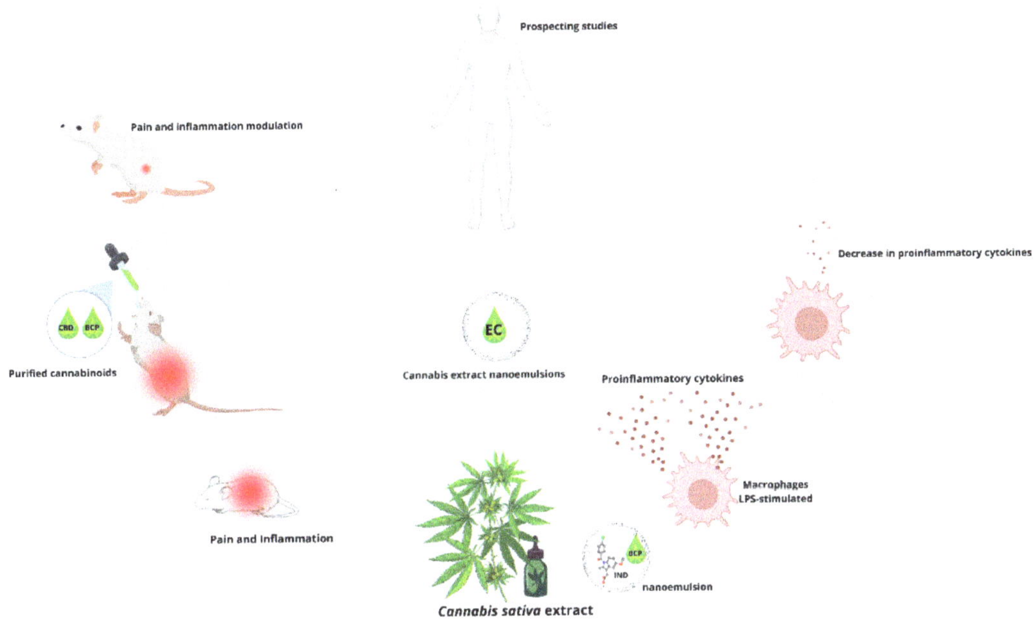

Figure 6. Immunomodulatory activity. The bioactive compounds cannabidiol (CBD) and caryophyllene (BCP) were administered orally in a murine model to treat pain and inflammation, reducing these conditions [108]. The mixture of indomethacin (IND) and BCP reduced the production of proinflammatory cytokines in a culture of macrophages stimulated by lipopolysaccharide (LPS) [109]. It is expected that studies with *Cannabis* extract nanoemulsions (NECs) will be carried out more frequently from now on.

7. Computational Analysis of *Cannabis sativa* Compounds

Computational analyses, in the context of *Cannabis sativa* bioactive compounds (isolated or in nanoemulsions), are necessary to understand their therapeutic properties better. Knowledge of these metabolites at the molecular level allows a better understanding of the interactions of these compounds with receptor proteins, helping to elucidate their mechanisms of action; for the precision of the metabolites in the *Cannabis* plant related to its genes and metabolic pathways, databases are used as proposed by [113]. Computational research has been carried out on the compounds provided by *Cannabis*. Researchers performed a characterization to explain the quantitative structure–activity relationships (QSARS) between cannabigerol (CBG), especially in the 3D structure, and the biological activity, complementing it with the density functional theory (DFT) to understand the stability and activity of the cannabinoid. In this way, they concluded that the geranyl chains of CBG follow a tendency to wrap around the central phenolic ring, leaving the side chains forming hydrogen bonds with the para-substituted hydroxyl groups. This research was also extended to molecular dockings with cytochrome P450-3A4, observing the decreased inhibitory effect of CBG on the key enzyme in drug metabolism, allowing a better understanding of the activity and structure of the compound [114].

Regarding immunomodulation, a study of bioactive peptides in *Cannabis* seed hydrolysates (bioHPHS) was carried out to evaluate their anti-inflammatory capacity in silico with the results obtained from in vitro tests in human monocytes [115]. Once the peptides in two bioHPHS were identified, ten were selected and synthesized in silico to determine their immunomodulatory capacity in monocytes. The peptides DDNPRRF, SRRFHLA, RNIFKGF, VREPVFSF, QADIFNPR, and SAERGFLY showed very high immunomodulatory activity. Subsequently, other molecular dockings were carried out to determine the interaction of the TLR4/MD2 peptide, seeking to understand the action in the inflammatory cascade modulated by these proteins. In a similar investigation, the impact of the cannabinoids CBD, CBG, and cannabinol (CBN) on DNA methylation was analyzed to understand the regulation of gene expression and cellular function; in this case, the target proteins were the TET enzymes important in DNA demethylation processes. Through computational simulations, it was shown that cannabinoids would have the ability to chelate ferrous ions, which could potentially interfere with the enzymatic activity, notably in CBD and CBN.

In an antimicrobial context, the antimalarial activity of compounds present in *Cannabis* was addressed in silico to elucidate in vivo results of tetrahydrocannabivarin (THCV) because this cannabinoid had a high affinity with the α/β tubulin proteins of the pathogen *Plasmodium falciparum* [116]. Based on these results, the authors suggested that this cannabinoid (THCV) would inhibit microtubules, preventing the parasite from advancing through red blood cells. In this way, one can understand the importance of computational analyses in the promising study of bioactive compounds of *C. sativa,* thanks to the fact that they allow a better understanding of the therapeutic properties and, at the same time, offer a broader view of the molecular interactions between these metabolites and the target receptors. Figure 7 is a graphical representation of this.

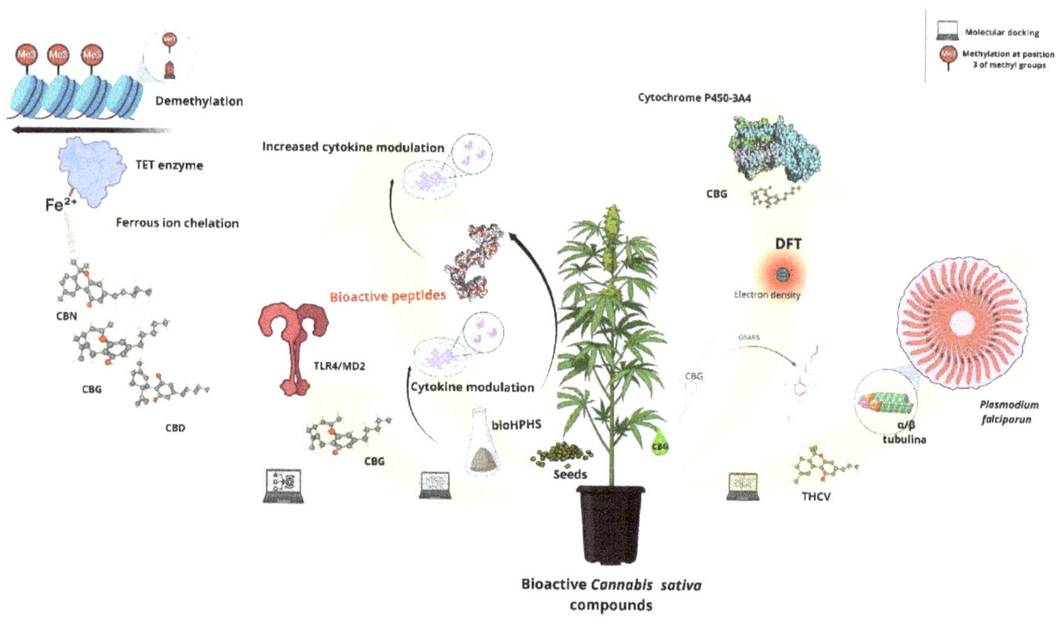

Figure 7. In silico studies. The analysis of computational studies on bioactive compounds from isolated metabolites and nanoemulsions of *Cannabis sativa* was parameterized in various functional modalities with comparative emphasis on therapeutic properties and molecular mechanisms. Such analyses include QSAR modeling, molecular docking, and DFT, which shed light on the degree of interaction between cannabinoids such as cannabigerol (CBG) with target proteins and their stabilities. One such example is CBG, whose inhibitory effects on cytochrome P450-3A4 were characterized [114]. Furthermore, some immunomodulatory peptides derived from *Cannabis* seed hydrolysates indicated some promising potential anti-inflammatory effects in silico and in vitro, especially since after the elucidation of the specific molecular mechanisms, they may prove useful as therapeutic agents [115]. Similarly, cannabinoids such as tetrahydrocannabivarin (THCV) were evaluated for their antimalarial activity by linking them to *Plasmodium falciparum* tubulin binding. These studies serve to underline the relevance of tools to explore the sources of bioactivity in *C. sativa* [116].

Although computational analysis exhibited some advantages, some challenges cannot be solved due to computational analytics requiring high data processing capacity and intensive operations [117]. This requires state-of-the-art equipment that integrates graphics processing units (GPUs) and central processing units (CPUs) well, ensuring efficient analyses. However, resource-limited academic contexts can limit this equipment's availability and cost. Once the appropriate hardware is configured, software with great analytical power is required; this includes data processing, visualization, and molecular modeling. These criteria constitute a challenge due to the variety, approach, learning curve, and costs of licenses [118]. In this context, artificial intelligence has gained space, and tools such as machine learning and data mining have proven to have great potential to detect patterns in data [119–121]; however, extensive knowledge of statistics, mathematics, and biological sciences is required. Because of these challenges, an interdisciplinary collaboration where mathematicians, physicists, chemists, engineers, and other specialties converge is necessary.

8. General Critical Points for *Cannabis sativa*-Based Nanosystems

Nanosystems have emerged as a technological proposal with wide application in biomedicine and food. These systems provide better active compound availability, distribution, and conservation [22]. However, *C. sativa*-based nanosystems seem underutilized in research, unlike purified cannabinoids, which have many applications and studies. This could impede the broad understanding of the therapeutic potential of the crude extract of the plant using nanoemulsions. The pharmaceutical industry could take part in the research now that discussions of legalizing the use of *Cannabis* are more frequent; this is convenient because it destigmatizes the use of the plant and supports the different applications of *Cannabis* in conjunction with biotechnology. In this way, the population could have more confidence in the therapeutic advantages and limitations of *Cannabis* nanoemulsions in various clinical contexts. The industry could participate in innovation in harnessing *Cannabis* and scaling up the production of nanoemulsions. Therefore, each new publication based on *Cannabis* nanoformulation should consider in vitro and in vivo release tests to extrapolate this technology, determining cytotoxicity, efficacy, and the best delivery method [122,123]. A key aspect is to evaluate the lipophilic nature of the extracts and convey them appropriately. From this point of view, lipid transporters in the formulations will be crucial for subsequent tests for both their release and toxicological properties [124]. In addition, regarding normative regulations, it is necessary to review the updates for human applications. New nanoformulations would be subject to scientific scrutiny and clinical trials approved by regulatory agencies. The regulatory framework differs from country to country, and the efforts of regulatory agencies such as the European Union and the FDA work in the face of the growing research based on cannabinoids and the current market demands.

The study of nanostructures, especially nanoemulsions, is highly anticipated. Although many of the techniques mentioned to produce nanoemulsions have not been applied, for the most part, in the study of EC, any advance in technology in terms of formulation, the combination of homogenization processes, or the improvement of interfacial properties could lead to much more stable and secure structures [125–128].

Finally, the possibilities of *C. sativa*-based nanosystems are vast, as cannabinoids are studied in neurosciences [129–132], cardiovascular health [133–135], gut motility [136–139], and inflammation [105].

9. Conclusions

C. sativa offers excellent therapeutic potential, including antimicrobial, immunomodulatory, and antioxidant properties. Aspects such as extraction and synthesis methods can determine the stability of nanostructures. However, despite all the benefits of the plant, studying it still presents challenges. Computational analyses make it easier to understand the properties of the compounds that comprise extracts and how they interact with molecular targets. Given all this, the pharmaceutical industry's active participation is needed to advance and exploit other forms of distribution in animals and, finally, application in humans.

Author Contributions: Conceptualization, C.A. and J.M.S.-J.; methodology, I.V., Z.Y.E.-G., R.C.B.-R., L.M.A.-E. and A.V.-L.; writing—original draft preparation, I.V., Z.Y.E.-G., R.C.B.-R., L.M.A.-E., A.V.-L., C.A. and J.M.S.-J.; writing—review and editing, I.V., C.A. and J.M.S.-J.; visualization, I.V., C.A. and J.M.S.-J.; supervision, C.A. and J.M.S.-J. All authors have read and agreed to the published version of the manuscript.

Funding: This research received no external funding.

Institutional Review Board Statement: Not applicable.

Informed Consent Statement: Not applicable.

Data Availability Statement: No new data were created or analyzed in this study. Data sharing is not applicable to this article.

Acknowledgments: The first author is grateful to the Consejo Nacional de Humanidades, Ciencias y Tecnologías (CONAHCyT) for the doctoral scholarship, CVU 1162015.

Conflicts of Interest: The authors declare no conflicts of interest.

References

1. Leal-Galicia, P.; Betancourt, D.; Gonzalez-Gonzalez, A.; Romo-Parra, H. A brief history of marijuana in the western world. *Rev. Neurol.* **2018**, *67*, 133–140. [PubMed]
2. Xie, Z.; Mi, Y.; Kong, L.; Gao, M.; Chen, S.; Chen, W.; Meng, X.; Sun, W.; Chen, S.; Xu, Z. *Cannabis sativa*: Origin and History, Glandular Trichome Development, and Cannabinoid Biosynthesis. *Hortic. Res.* **2023**, *10*, 15. [CrossRef] [PubMed]
3. Livingston, S.J.; Quilichini, T.D.; Booth, J.K.; Wong, D.C.J.; Rensing, K.H.; Laflamme-Yonkman, J.; Castellarin, S.D.; Bohlmann, J.; Page, J.E.; Samuels, A.L. Cannabis Glandular Trichomes Alter Morphology and Metabolite Content during Flower Maturation. *Plant J.* **2020**, *101*, 37–56. [CrossRef]
4. López Carretero, P.; Pekas, A.; Stubsgaard, L.; Sancho Blanco, G.; Lütken, H.; Sigsgaard, L. Glandular Trichomes Affect Mobility and Predatory Behavior of Two Aphid Predators on Medicinal Cannabis. *Biol. Control* **2022**, *170*, 104932. [CrossRef]
5. Hourfane, S.; Mechqoq, H.; Bekkali, A.Y.; Rocha, J.M.; El Aouad, N. A Comprehensive Review on *Cannabis sativa* Ethnobotany, Phytochemistry, Molecular Docking and Biological Activities. *Plants* **2023**, *12*, 1245. [CrossRef]
6. Salamone, S.; Waltl, L.; Pompignan, A.; Grassi, G.; Chianese, G.; Koeberle, A.; Pollastro, F. Phytochemical Characterization of *Cannabis sativa* L. Chemotype V Reveals Three New Dihydrophenanthrenoids That Favorably Reprogram Lipid Mediator Biosynthesis in Macrophages. *Plants* **2022**, *11*, 2130. [CrossRef]
7. Rodríguez Mesa, X.M.; Moreno Vergara, A.F.; Contreras Bolaños, L.A.; Guevara Moriones, N.; Mejía Piñeros, A.L.; Santander González, S.P. Therapeutic Prospects of Cannabinoids in the Immunomodulation of Prevalent Autoimmune Diseases. *Cannabis Cannabinoid Res.* **2021**, *6*, 196–210. [CrossRef]
8. Kubiliene, A.; Mickute, K.; Baranauskaite, J.; Marksa, M.; Liekis, A.; Sadauskiene, I. The Effects of *Cannabis sativa* L. Extract on Oxidative Stress Markers In Vivo. *Life* **2021**, *11*, 647. [CrossRef]
9. Ali, E.M.M.; Almagboul, A.Z.I.; Khogali, S.M.E.; Gergeir, U.M.A. Antimicrobial Activity of *Cannabis sativa* L. *Chin. Med.* **2012**, *3*, 61–64. [CrossRef]
10. Flores-Sanchez, I.J.; Verpoorte, R. Secondary Metabolism in Cannabis. *Phytochem. Rev.* **2008**, *7*, 615–639. [CrossRef]
11. Karas, J.A.; Wong, L.J.M.; Paulin, O.K.A.; Mazeh, A.C.; Hussein, M.H.; Li, J.; Velkov, T. The Antimicrobial Activity of Cannabinoids. *Antibiotics* **2020**, *9*, 406. [CrossRef] [PubMed]
12. Leinen, Z.J.; Mohan, R.; Premadasa, L.S.; Acharya, A.; Mohan, M.; Byrareddy, S.N. Therapeutic Potential of Cannabis: A Comprehensive Review of Current and Future Applications. *Biomedicines* **2023**, *11*, 2630. [CrossRef]
13. Nichols, J.M.; Kummari, E.; Sherman, J.; Yang, E.-J.; Dhital, S.; Gilfeather, C.; Yray, G.; Morgan, T.; Kaplan, B.L.F. CBD Suppression of EAE Is Correlated with Early Inhibition of Splenic IFN-γ + CD8+ T Cells and Modest Inhibition of Neuroinflammation. *J. Neuroimmune Pharmacol.* **2021**, *16*, 346–362. [CrossRef]
14. Ben-Shabat, S.; Fride, E.; Sheskin, T.; Tamiri, T.; Rhee, M.H.; Vogel, Z.; Bisogno, T.; De Petrocellis, L.; Di Marzo, V.; Mechoulam, R. An Entourage Effect: Inactive Endogenous Fatty Acid Glycerol Esters Enhance 2-Arachidonoyl-Glycerol Cannabinoid Activity. *Eur. J. Pharmacol.* **1998**, *353*, 23–31. [CrossRef]
15. Opęchowska, A.; Karpiuk, K.; Zahorodnii, A.; Harasim-Symbor, E.; Chabowski, A.; Konstantynowicz-Nowicka, K. Anti-Inflammatory Effects of Cannabidiol in Early Stages of Neuroinflammation Induced by High-Fat Diet in Cerebral Cortex of Rats. *Toxicol. Appl. Pharmacol.* **2024**, *484*, 116856. [CrossRef] [PubMed]

16. Jagannathan, R. Identification of Psychoactive Metabolites from *Cannabis sativa*, Its Smoke, and Other Phytocannabinoids Using Machine Learning and Multivariate Methods. *ACS Omega* **2020**, *5*, 281–295. [CrossRef] [PubMed]
17. Duong, V.-A.; Nguyen, T.-T.-L.; Maeng, H.-J. Preparation of Solid Lipid Nanoparticles and Nanostructured Lipid Carriers for Drug Delivery and the Effects of Preparation Parameters of Solvent Injection Method. *Molecules* **2020**, *25*, 4781. [CrossRef]
18. Nair, A.; Mallya, R.; Suvarna, V.; Khan, T.A.; Momin, M.; Omri, A. Nanoparticles—Attractive Carriers of Antimicrobial Essential Oils. *Antibiotics* **2022**, *11*, 108. [CrossRef]
19. Chayasirisobhon, S. Mechanisms of Action and Pharmacokinetics of Cannabis. *Perm. J.* **2020**, *25*, 19.200. [CrossRef]
20. Martínez, V.; Iriondo De-Hond, A.; Borrelli, F.; Capasso, R.; del Castillo, M.D.; Abalo, R. Cannabidiol and Other Non-Psychoactive Cannabinoids for Prevention and Treatment of Gastrointestinal Disorders: Useful Nutraceuticals? *Int. J. Mol. Sci.* **2020**, *21*, 3067. [CrossRef]
21. Abreu, F.O.M.S.; Oliveira, E.F.; Paula, H.C.B.; de Paula, R.C.M. Chitosan/Cashew Gum Nanogels for Essential Oil Encapsulation. *Carbohydr. Polym.* **2012**, *89*, 1277–1282. [CrossRef] [PubMed]
22. Choi, S.J.; McClements, D.J. Nanoemulsions as Delivery Systems for Lipophilic Nutraceuticals: Strategies for Improving Their Formulation, Stability, Functionality and Bioavailability. *Food Sci. Biotechnol.* **2020**, *29*, 149–168. [CrossRef] [PubMed]
23. Cleirec, G.; Desmier, E.; Lacatus, C.; Lesgourgues, S.; Braun, A.; Peloso, C.; Obadia, C. Efficiency of Inhaled Cannabidiol in Cannabis Use Disorder: The Pilot Study Cannavap. *Front. Psychiatry* **2022**, *13*, 899221. [CrossRef]
24. Lucas, C.J.; Galettis, P.; Schneider, J. The Pharmacokinetics and the Pharmacodynamics of Cannabinoids. *Br. J. Clin. Pharmacol.* **2018**, *84*, 2477–2482. [CrossRef] [PubMed]
25. Arianto, A.; Cindy, C. Preparation and Evaluation of Sunflower Oil Nanoemulsion as a Sunscreen. *Open Access Maced. J. Med. Sci.* **2019**, *7*, 3757–3761. [CrossRef]
26. Ren, J.-N.; Dong, M.; Hou, Y.-Y.; Fan, G.; Pan, S.-Y. Effect of Olive Oil on the Preparation of Nanoemulsions and Its Effect on Aroma Release. *J. Food Sci. Technol.* **2018**, *55*, 4223–4231. [CrossRef]
27. He, W.; Tan, Y.; Tian, Z.; Chen, L.; Hu, F.; Wu, W. Food Protein-Stabilized Nanoemulsions as Potential Delivery Systems for Poorly Water-Soluble Drugs: Preparation, in Vitro Characterization, and Pharmacokinetics in Rats. *Int. J. Nanomed.* **2011**, *6*, 521–533. [CrossRef]
28. Kandadi, P.; Syed, M.A.; Goparaboina, S.; Veerabrahma, K. Albumin Coupled Lipid Nanoemulsions of Diclofenac for Targeted Delivery to Inflammation. *Nanomedicine* **2012**, *8*, 1162–1171. [CrossRef]
29. Li, Y.; Jin, H.; Sun, X.; Sun, J.; Liu, C.; Liu, C.; Xu, J. Physicochemical Properties and Storage Stability of Food Protein-Stabilized Nanoemulsions. *Nanomaterials* **2018**, *9*, 25. [CrossRef]
30. Espinosa-Sandoval, L.; Ochoa-Martínez, C.; Ayala-Aponte, A.; Pastrana, L.; Gonçalves, C.; Cerqueira, M.A. Polysaccharide-Based Multilayer Nano-Emulsions Loaded with Oregano Oil: Production, Characterization, and In Vitro Digestion Assessment. *Nanomaterials* **2021**, *11*, 878. [CrossRef]
31. Malik, M.R.; Al-Harbi, F.F.; Nawaz, A.; Amin, A.; Farid, A.; Mohaini, M.A.; Alsalman, A.J.; Hawaj, M.A.A.; Alhashem, Y.N. Formulation and Characterization of Chitosan-Decorated Multiple Nanoemulsion for Topical Delivery In Vitro and Ex Vivo. *Molecules* **2022**, *27*, 3183. [CrossRef] [PubMed]
32. Qiu, A.; Wang, Y.; Zhang, G.; Wang, H. Natural Polysaccharide-Based Nanodrug Delivery Systems for Treatment of Diabetes. *Polymers* **2022**, *14*, 3217. [CrossRef] [PubMed]
33. Roda Zitha Vilanculos, J.; Silva de Farias, B.; Inês Engelmann, J.; Silveira Ribeiro, E.; Diaz de Oliveira, P.; Roberto Sant'Anna Cadaval, T.; Antonio de Almeida Pinto, L. Physicochemical Evaluation of Chitosan–Xanthan Gum Nanoemulsions as Polyunsaturated Enriched Lipid–Carrier. *J. Mol. Liq.* **2023**, *386*, 122533. [CrossRef]
34. Wang, L.; Zhan, J.; Ma, R.; Tian, Y. Preparation of Starch-Based Nanoemulsion for Sustained Release and Enhanced Bioaccessibility of Quercetin. *Colloids Surf. A Physicochem. Eng. Asp.* **2023**, *665*, 131218. [CrossRef]
35. Himanath, G.; Shruthy, R.; Preetha, R.; Sreejit, V. Nanoemulsion with Coconut Oil and Soy Lecithin as a Stable Delivery System for Lycopene and Its Incorporation into Yogurt to Enhance Antioxidant Properties and Maintain Quality. *ACS Food Sci. Technol.* **2021**, *1*, 1538–1549. [CrossRef]
36. Mushtaq, A.; Mohd Wani, S.; Malik, A.R.; Gull, A.; Ramniwas, S.; Ahmad Nayik, G.; Ercisli, S.; Alina Marc, R.; Ullah, R.; Bari, A. Recent Insights into Nanoemulsions: Their Preparation, Properties and Applications. *Food Chem. X* **2023**, *18*, 100684. [CrossRef]
37. Schreiner, T.B.; Santamaria-Echart, A.; Ribeiro, A.; Peres, A.M.; Dias, M.M.; Pinho, S.P.; Barreiro, M.F. Formulation and Optimization of Nanoemulsions Using the Natural Surfactant Saponin from Quillaja Bark. *Molecules* **2020**, *25*, 1538. [CrossRef]
38. Rožanc, J.; Kotnik, P.; Milojević, M.; Gradišnik, L.; Knez Hrnčič, M.; Knez, Ž.; Maver, U. Different *Cannabis sativa* Extraction Methods Result in Different Biological Activities against a Colon Cancer Cell Line and Healthy Colon Cells. *Plants* **2021**, *10*, 566. [CrossRef]
39. Blake, A.; Nahtigal, I. The Evolving Landscape of Cannabis Edibles. *Curr. Opin. Food Sci.* **2019**, *28*, 25–31. [CrossRef]
40. Fischedick, J.T.; Hazekamp, A.; Erkelens, T.; Choi, Y.H.; Verpoorte, R. Metabolic Fingerprinting of *Cannabis sativa* L., Cannabinoids and Terpenoids for Chemotaxonomic and Drug Standardization Purposes. *Phytochemistry* **2010**, *71*, 2058–2073. [CrossRef]

41. Radoiu, M.; Kaur, H.; Bakowska-Barczak, A.; Splinter, S. Microwave-Assisted Industrial Scale Cannabis Extraction. *Technologies* **2020**, *8*, 45. [CrossRef]
42. Dussy, F.E.; Hamberg, C.; Luginbühl, M.; Schwerzmann, T.; Briellmann, T.A. Isolation of Delta9-THCA-A from Hemp and Analytical Aspects Concerning the Determination of Delta9-THC in Cannabis Products. *Forensic Sci. Int.* **2005**, *149*, 3–10 [CrossRef] [PubMed]
43. Tiago, F.J.; Paiva, A.; Matias, A.A.; Duarte, A.R.C. Extraction of Bioactive Compounds from *Cannabis sativa* L. Flowers and/or Leaves Using Deep Eutectic Solvents. *Front. Nutr.* **2022**, *9*, 892314. [CrossRef]
44. Madia, V.N.; Di Santo, R.; Costi, R. Chapter 2—Medical Cannabis and Cannabinoids: How Best to Extract Components from Plant Material. In *Medicinal Usage of Cannabis and Cannabinoids*; Preedy, V.R., Patel, V.B., Martin, C.R., Eds.; Academic Press: Cambridge, MA, USA, 2023; pp. 15–23, ISBN 978-0-323-90036-2.
45. Sungpud, C.; Panpipat, W.; Sae Yoon, A.; Chaijan, M. Ultrasonic-Assisted Virgin Coconut Oil Based Extraction for Maximizing Polyphenol Recovery and Bioactivities of Mangosteen Peels. *J. Food Sci. Technol.* **2020**, *57*, 4032–4043. [CrossRef] [PubMed]
46. Sousa, G.; Trifunovska, M.; Antunes, M.; Miranda, I.; Moldão, M.; Alves, V.; Vidrih, R.; Lopes, P.A.; Aparicio, L.; Neves, M.; et al Optimization of Ultrasound-Assisted Extraction of Bioactive Compounds from *Pelvetia canaliculata* to Sunflower Oil. *Foods* **2021**, *10*, 1732. [CrossRef]
47. Casiraghi, A.; Roda, G.; Casagni, E.; Cristina, C.; Musazzi, U.M.; Franzè, S.; Rocco, P.; Giuliani, C.; Fico, G.; Minghetti, P.; et al Extraction Method and Analysis of Cannabinoids in Cannabis Olive Oil Preparations. *Planta Med.* **2018**, *84*, 242–249. [CrossRef]
48. Cravotto, C.; Fabiano-Tixier, A.-S.; Claux, O.; Abert-Vian, M.; Tabasso, S.; Cravotto, G.; Chemat, F. Towards Substitution of Hexane as Extraction Solvent of Food Products and Ingredients with No Regrets. *Foods* **2022**, *11*, 3412. [CrossRef]
49. Koina, I.M.; Sarigiannis, Y.; Hapeshi, E. Green Extraction Techniques for the Determination of Active Ingredients in Tea: Current State, Challenges, and Future Perspectives. *Separations* **2023**, *10*, 121. [CrossRef]
50. Discover Advanced Hydrodynamic Cannabis Extraction—Cannabis Tech. 2018. Available online: https://cannabistech.com/articles/hydrodynamic-cannabis-extraction/ (accessed on 5 March 2024).
51. Carabias-Martínez, R.; Rodríguez-Gonzalo, E.; Revilla-Ruiz, P.; Hernández-Méndez, J. Pressurized Liquid Extraction in the Analysis of Food and Biological Samples. *J. Chromatogr. A* **2005**, *1089*, 1–17. [CrossRef]
52. Bobinaitė, R.; Kraujalis, P.; Tamkutė, L.; Urbonavičienė, D.; Viškelis, P.; Venskutonis, P.R. Recovery of Bioactive Substances from Rowanberry Pomace by Consecutive Extraction with Supercritical Carbon Dioxide and Pressurized Solvents. *J. Ind. Eng. Chem.* **2020**, *85*, 152–160. [CrossRef]
53. Pilařová, V.; Hadysová, Z.; Švec, F.; Nováková, L. Supercritical Fluids in Analysis of Cannabinoids in Various Cannabis Products. *Anal. Chim. Acta* **2022**, *1232*, 340452. [CrossRef] [PubMed]
54. Jafari, S.; McClements, D. *Nanoemulsions: Formulation, Applications, and Characterization*; Academic Press: Cambridge, MA, USA, 2018.
55. Addo, P.W.; Sagili, S.U.K.R.; Bilodeau, S.E.; Gladu-Gallant, F.-A.; MacKenzie, D.A.; Bates, J.; McRae, G.; MacPherson, S.; Paris, M.; Raghavan, V.; et al. Microwave- and Ultrasound-Assisted Extraction of Cannabinoids and Terpenes from Cannabis Using Response Surface Methodology. *Molecules* **2022**, *27*, 8803. [CrossRef] [PubMed]
56. Lazarjani, M.P.; Young, O.; Kebede, L.; Seyfoddin, A. Processing and Extraction Methods of Medicinal Cannabis: A Narrative Review. *J. Cannabis Res.* **2021**, *3*, 32. [CrossRef] [PubMed]
57. Buitrago, O.Y.; Ardila, R.; Orjuela, A.; Santaella, M.A.; Arturo, D.E.; Hurtado, A. Affordable Method for Batch Supercritical Extraction Using Solid Carbon Dioxide–Extraction of Cannabis Threshing Residues. *Chem. Eng. Process.—Process Intensif.* **2024**, *198*, 109721. [CrossRef]
58. Naveira-Pazos, C.; Veiga, M.C.; Mussagy, C.U.; Farias, F.O.; Kennes, C.; Pereira, J.F.B. Carotenoids Production and Extraction from Yarrowia Lipolytica Cells: A Biocompatible Approach Using Biosolvents. *Sep. Purif. Technol.* **2024**, *343*, 127136. [CrossRef]
59. Han, Q.-H.; Liu, W.; Li, H.-Y.; He, J.-L.; Guo, H.; Lin, S.; Zhao, L.; Chen, H.; Liu, Y.-W.; Wu, D.-T.; et al. Extraction Optimization, Physicochemical Characteristics, and Antioxidant Activities of Polysaccharides from Kiwifruit (*Actinidia chinensis* Planch.). *Molecules* **2019**, *24*, 461. [CrossRef]
60. Chouhan, K.B.S.; Tandey, R.; Sen, K.K.; Mehta, R.; Mandal, V. A Unique Model of Gravity Assisted Solvent Free Microwave Based Extraction of Essential Oil from Mentha Leaves Ensuring Biorefinery of Leftover Waste Biomass for Extraction of Nutraceuticals: Towards Cleaner and Greener Technology. *J. Clean. Prod.* **2019**, *225*, 587–598. [CrossRef]
61. Sorita, G.D.; Favaro, S.P.; Ambrosi, A.; Luccio, M.D. Aqueous Extraction Processing: An Innovative and Sustainable Approach for Recovery of Unconventional Oils. *Trends Food Sci. Technol.* **2023**, *133*, 99–113. [CrossRef]
62. McClements, D.J. *Food Emulsions: Principles, Practices, and Techniques*, 2nd ed.; CRC Press: Boca Raton, FL, USA, 2004; ISBN 978-0-429-12389-4.
63. Gan, L.; Cui, D.; Ali, N.; Zhang, Q.; Zhang, D.; Jiang, W.; Zhang, W. Phase Behavior of Polyglycerol Ester-Based Nanoemulsions. *J. Nanosci. Nanotechnol.* **2021**, *21*, 6188–6195. [CrossRef]
64. Zeeb, B.; Herz, E.; McClements, D.; Weiss, J. Impact of Alcohols on the Formation and Stability of Protein-Stabilized Nanoemulsions. *J. Colloid. Interface Sci.* **2014**, *433*, 196–203. [CrossRef]

65. Bleoanca, I.; Lanciu, A.; Patrașcu, L.; Ceoromila, A.; Borda, D. Efficacy of Two Stabilizers in Nanoemulsions with Whey Proteins and Thyme Essential Oil as Edible Coatings for Zucchini. *Membranes* **2022**, *12*, 326. [CrossRef] [PubMed]
66. Abbasi, Z.; Aminzare, M.; Hassanzad Azar, H.; Rostamizadeh, K. Effect of Corn Starch Coating Incorporated with Nanoemulsion of Zataria Multiflora Essential Oil Fortified with Cinnamaldehyde on Microbial Quality of Fresh Chicken Meat and Fate of Inoculated Listeria Monocytogenes. *J. Food Sci. Technol.* **2021**, *58*, 2677–2687. [CrossRef] [PubMed]
67. Agame-Lagunes, B.; Grube-Pagola, P.; García-Varela, R.; Alexander-Aguilera, A.; García, H.S. Effect of Curcumin Nanoemulsions Stabilized with MAG and DAG-MCFAs in a Fructose-Induced Hepatic Steatosis Rat Model. *Pharmaceutics* **2021**, *13*, 509. [CrossRef]
68. Vega-Hernández, L.C.; Serrano-Niño, J.C.; Velázquez-Carriles, C.A.; Martínez-Preciado, A.H.; Cavazos-Garduño, A.; Silva-Jara, J.M. Improving Foodborne Pathogen Control Using Green Nanosized Emulsions of *Plectranthus hadiensis* Phytochemicals. *Colloids Interfaces* **2024**, *8*, 3. [CrossRef]
69. Kim, T.-I.; Kim, T.-G.; Lim, D.-H.; Kim, S.-B.; Park, S.-M.; Hur, T.-Y.; Ki, K.-S.; Kwon, E.-G.; Vijayakumar, M.; Kim, Y.-J. Preparation of Nanoemulsions of Vitamin A and C by Microfluidization: Efficacy on the Expression Pattern of Milk-Specific Proteins in MAC-T Cells. *Molecules* **2019**, *24*, 2566. [CrossRef]
70. Knoke, S.; Bunjes, H. Transfer of Lipophilic Drugs from Nanoemulsions into Lipid-Containing Alginate Microspheres. *Pharmaceutics* **2021**, *13*, 173. [CrossRef]
71. Vater, C.; Bosch, L.; Mitter, A.; Göls, T.; Seiser, S.; Heiss, E.; Elbe-Bürger, A.; Wirth, M.; Valenta, C.; Klang, V. Lecithin-Based Nanoemulsions of Traditional Herbal Wound Healing Agents and Their Effect on Human Skin Cells. *Eur. J. Pharm. Biopharm.* **2022**, *170*, 1–9. [CrossRef]
72. Ali, A.; Mekhloufi, G.; Huang, N.; Agnely, F. β-Lactoglobulin Stabilized Nanemulsions--Formulation and Process Factors Affecting Droplet Size and Nanoemulsion Stability. *Int. J. Pharm.* **2016**, *500*, 291–304. [CrossRef] [PubMed]
73. Abbasian Chaleshtari, Z.; Salimi-Kenari, H.; Foudazi, R. Glassy and Compressed Nanoemulsions Stabilized with Sodium Dodecyl Sulfate in the Presence of Poly(Ethylene Glycol)-Diacrylate. *Soft Matter* **2023**, *19*, 5989–6004. [CrossRef]
74. Azeem, A.; Rizwan, M.; Ahmad, F.J.; Iqbal, Z.; Khar, R.K.; Aqil, M.; Talegaonkar, S. Nanoemulsion Components Screening and Selection: A Technical Note. *AAPS PharmSciTech* **2009**, *10*, 69–76. [CrossRef]
75. Algahtani, M.S.; Ahmad, M.Z.; Ahmad, J. Investigation of Factors Influencing Formation of Nanoemulsion by Spontaneous Emulsification: Impact on Droplet Size, Polydispersity Index, and Stability. *Bioengineering* **2022**, *9*, 384. [CrossRef] [PubMed]
76. Zhang, M.; Cao, Q.; Yuan, Y.; Guo, X.; Pan, D.; Xie, R.; Ju, X.; Liu, Z.; Wang, W.; Chu, L. Functional Nanoemulsions: Controllable Low-Energy Nanoemulsification and Advanced Biomedical Application. *Chin. Chem. Lett.* **2024**, *35*, 108710. [CrossRef]
77. Hadžiabdić, J.; Džana, O.; Vranić, E.; Ognjenka, R. Preparation of nanoemulsions by high-energy and lowenergy emulsification methods. In *IFMBE Proceedings*; Springer: Berlin/Heidelberg, Germany, 2017; pp. 317–322.
78. Song, R.; Lin, Y.; Li, Z. Ultrasonic-Assisted Preparation of Eucalyptus Oil Nanoemulsion: Process Optimization, In Vitro Digestive Stability, and Anti-Escherichia Coli Activity. *Ultrason. Sonochem.* **2021**, *82*, 105904. [CrossRef]
79. Ochoa, A.A.; Hernández-Becerra, J.A.; Cavazos-Garduño, A.; Vernon-Carter, E.J.; García, H.S.; Ochoa, A.A.; Hernández-Becerra, J.A.; Cavazos-Garduño, A.; Vernon-Carter, E.J.; García, H.S. Preparation and Characterization of Curcumin Nanoemulsions Obtained by Thin-Film Hydration Emulsification and Ultrasonication Methods. *Rev. Mex. Ing. Química* **2016**, *15*, 79–90.
80. Sharma, S.; Sahni, J.K.; Ali, J.; Baboota, S. Effect of High-Pressure Homogenization on Formulation of TPGS Loaded Nanoemulsion of Rutin—Pharmacodynamic and Antioxidant Studies. *Drug Deliv.* **2015**, *22*, 541–551. [CrossRef] [PubMed]
81. Guerra Rosas, M.I.; Morales-Castro, J.; Ochoa-Martínez, L.A.; Salvia-Trujillo, L.; Martín-Belloso, O. Long-Term Stability of Food-Grade Nanoemulsions from High Methoxyl Pectin Containing Essential Oils. *Food Hydrocoll.* **2016**, *52*, 438–446. [CrossRef]
82. Danhier, F.; Ansorena, E.; Silva, J.; Coco, R.; Aude, L.B.; Préat, V. PLGA-Based Nanoparticles: An Overview of Biomedical Applications. *J. Control. Release* **2012**, *161*, 505–522. [CrossRef] [PubMed]
83. Joseph, S. Bunjes Preparation of Nanoemulsions and Solid Lipid Nanoparticles by Premix Membrane Emulsification. *J. Pharm. Sci.* **2012**, *101*, 2479–2489. [CrossRef] [PubMed]
84. Kaminski, A.; McBain, J.W. Spontaneous Emulsification of Pure Xylene in an Aqueous Solution through Mere Adsorption of a Detergent in the Interface. *Proc. R. Soc. Lond. Ser. A Math. Phys. Sci.* **1997**, *198*, 447–454. [CrossRef]
85. McBain, J.W.; Woo, T. Spontaneous Emulsification, and Reactions Overshooting Equilibrium. *Proc. R. Soc. Lond. Ser. A Math. Phys. Sci.* **1997**, *163*, 182–188. [CrossRef]
86. Kedare, S.B.; Singh, R.P. Genesis and Development of DPPH Method of Antioxidant Assay. *J. Food Sci. Technol.* **2011**, *48*, 412–422. [CrossRef] [PubMed]
87. Kumar, H.; Bhardwaj, K.; Nepovimova, E.; Kuča, K.; Dhanjal, D.S.; Bhardwaj, S.; Bhatia, S.K.; Verma, R.; Kumar, D. Antioxidant Functionalized Nanoparticles: A Combat against Oxidative Stress. *Nanomaterials* **2020**, *10*, 1334. [CrossRef] [PubMed]
88. Dos-Santos-Pereira, M.; Guimarães, F.S.; Del-Bel, E.; Raisman-Vozari, R.; Michel, P.P. Cannabidiol Prevents LPS-Induced Microglial Inflammation by Inhibiting ROS/NF-κB-Dependent Signaling and Glucose Consumption. *Glia* **2020**, *68*, 561–573. [CrossRef]
89. Raja, A.; Ahmadi, S.; de Costa, F.; Li, N.; Kerman, K. Attenuation of Oxidative Stress by Cannabinoids and Cannabis Extracts in Differentiated Neuronal Cells. *Pharmaceuticals* **2020**, *13*, 328. [CrossRef]
90. Pollastro, F.; Minassi, A.; Fresu, L.G. Cannabis Phenolics and Their Bioactivities. *Curr. Med. Chem.* **2018**, *25*, 1160–1185. [CrossRef]

91. Kopustinskiene, D.M.; Masteikova, R.; Lazauskas, R.; Bernatoniene, J. *Cannabis sativa* L. Bioactive Compounds and Their Protective Role in Oxidative Stress and Inflammation. *Antioxidants* **2022**, *11*, 660. [CrossRef]
92. Berardo, M.E.V.; Mendieta, J.R.; Villamonte, M.D.; Colman, S.L.; Nercessian, D. Antifungal and Antibacterial Activities of *Cannabis sativa* L. Resins. *J. Ethnopharmacol.* **2024**, *318*, 116839. [CrossRef]
93. Demisli, S.; Galani, E.; Goulielmaki, M.; Kyrilis, F.L.; Ilić, T.; Hamdi, F.; Crevar, M.; Kastritis, P.L.; Pletsa, V.; Nallet, F.; et al. Encapsulation of Cannabidiol in Oil-in-Water Nanoemulsions and Nanoemulsion-Filled Hydrogels: A Structure and Biological Assessment Study. *J. Colloid Interface Sci.* **2023**, *634*, 300–313. [CrossRef] [PubMed]
94. Pugazhendhi, A.; Suganthy, N.; Chau, T.P.; Sharma, A.; Unpaprom, Y.; Ramaraj, R.; Karuppusamy, I.; Brindhadevi, K. Cannabinoids as Anticancer and Neuroprotective Drugs: Structural Insights and Pharmacological Interactions—A Review. *Process Biochem.* **2021**, *111*, 9–31. [CrossRef]
95. Rebibo, L.; Frušić-Zlotkin, M.; Ofri, R.; Nassar, T.; Benita, S. The Dose-Dependent Effect of a Stabilized Cannabidiol Nanoemulsion on Ocular Surface Inflammation and Intraocular Pressure. *Int. J. Pharm.* **2022**, *617*, 121627. [CrossRef]
96. Tran, V.N.; Strnad, O.; Šuman, J.; Veverková, T.; Sukupová, A.; Cejnar, P.; Hynek, R.; Kronusová, O.; Šach, J.; Kaštánek, P.; et al. Cannabidiol Nanoemulsion for Eye Treatment—Anti-Inflammatory, Wound Healing Activity and Its Bioavailability Using In Vitro Human Corneal Substitute. *Int. J. Pharm.* **2023**, *643*, 123202. [CrossRef]
97. Pagano, C.; Savarese, B.; Coppola, L.; Navarra, G.; Avilia, G.; Laezza, C.; Bifulco, M. Cannabinoids in the Modulation of Oxidative Signaling. *Int. J. Mol. Sci.* **2023**, *24*, 2513. [CrossRef]
98. Frassinetti, S.; Gabriele, M.; Moccia, E.; Longo, V.; Gioia, D.D. Antimicrobial and Antibiofilm Activity of *Cannabis sativa* L. Seeds Extract against Staphylococcus Aureus and Growth Effects on Probiotic *Lactobacillus* spp. *LWT* **2020**, *124*, 109149. [CrossRef]
99. Elhendawy, M.A.; Wanas, A.S.; Radwan, M.M.; Azzaz, N.A.; Toson, E.S.; ElSohly, M.A. Chemical and Biological Studies of *Cannabis sativa* Roots. *Med. Cannabis Cannabinoids* **2019**, *1*, 104–111. [CrossRef]
100. Zengin, G.; Menghini, L.; Di Sotto, A.; Mancinelli, R.; Sisto, F.; Carradori, S.; Cesa, S.; Fraschetti, C.; Filippi, A.; Angiolella, L.; et al. Chromatographic Analyses, In Vitro Biological Activities, and Cytotoxicity of *Cannabis sativa* L. Essential Oil: A Multidisciplinary Study. *Molecules* **2018**, *23*, 3266. [CrossRef]
101. Menossi, M.; Tejada, G.; Colman, S.L.; Nercessian, D.; Mendieta, J.R.; Islan, G.A.; Alvarez, V.A. Cannabis Extract-Loaded Lipid and Chitosan-Coated Lipid Nanoparticles with Antifungal Activity. *Colloids Surf. A Physicochem. Eng. Asp.* **2024**, *685*, 133207. [CrossRef]
102. Michailidu, J.; Miškovská, A.; Jarošová, I.; Čejková, A.; Mat'átková, O. Bimetallic Nanoparticle Production Using *Cannabis sativa* and Vitis Vinifera Waste Extracts. *RSC Adv.* **2024**, *14*, 5309–5318. [CrossRef]
103. Friedman, H.; Klein, T.W.; Newton, C.; Daaka, Y. Marijuana, Receptors and Immunomodulation. *Adv. Exp. Med. Biol.* **1995**, *373*, 103–113. [CrossRef]
104. Blevins, L.K.; Bach, A.P.; Crawford, R.B.; Zhou, J.; Henriquez, J.E.; Rizzo, M.D.; Sermet, S.; Khan, D.M.I.O.; Turner, H.; Small-Howard, A.L.; et al. Evaluation of the Anti-Inflammatory Effects of Selected Cannabinoids and Terpenes from *Cannabis sativa* Employing Human Primary Leukocytes. *Food Chem. Toxicol.* **2022**, *170*, 113458. [CrossRef]
105. Nayak, A.P.; Loblundo, C.; Bielory, L. Immunomodulatory Actions of Cannabinoids: Clinical Correlates and Therapeutic Opportunities for Allergic Inflammation. *J. Allergy Clin. Immunol. Pr.* **2023**, *11*, 449–457. [CrossRef] [PubMed]
106. Stella, N. THC and CBD: Similarities and Differences between Siblings. *Neuron* **2023**, *111*, 302–327. [CrossRef] [PubMed]
107. Blanton, H.; Yin, L.; Duong, J.; Benamar, K. Cannabidiol and Beta-Caryophyllene in Combination: A Therapeutic Functional Interaction. *Int. J. Mol. Sci.* **2022**, *23*, 15470. [CrossRef]
108. De Oliveira Carvalho, H.; de Melo Santos, A.; de Lima Teixeira dos Santos, A.V.T.; Gonçalves, D.E.S.; Picanço, K.R.T.; Souza, B.S.F.e.; Carvalho, J.C.T. *Cannabis sativa* L. Fixed Oil and Its Nanoemulsion: Effect on Diabetes and Dyslipidemia Induced in Rats. *Pharmacogn. Mag.* **2024**, *20*, 908–920. [CrossRef]
109. Weimer, P.; Kirsten, C.N.; de Araújo Lock, G.; Nunes, K.A.A.; Rossi, R.C.; Koester, L.S. Co-Delivery of Beta-Caryophyllene and Indomethacin in the Oily Core of Nanoemulsions Potentiates the Anti-Inflammatory Effect in LPS-Stimulated Macrophage Model. *Eur. J. Pharm. Biopharm.* **2023**, *191*, 114–123. [CrossRef]
110. Acharya, N.; Penukonda, S.; Shcheglova, T.; Hagymasi, A.T.; Basu, S.; Srivastava, P.K. Endocannabinoid System Acts as a Regulator of Immune Homeostasis in the Gut. *Proc. Natl. Acad. Sci. USA* **2017**, *114*, 5005–5010. [CrossRef]
111. Brown, K.J.; Laun, A.S.; Song, Z.-H. Cannabidiol, a Novel Inverse Agonist for GPR12. *Biochem. Biophys. Res. Commun.* **2017**, *493*, 451–454. [CrossRef]
112. Murphy, K.; Mowat, A.; Weaver, C. *Janeway's Immunobiology*; Garland Science, Taylor & Francis Group: New York, NY, USA, 2012; ISBN 978-0-8153-4243-4.
113. Cai, S.; Zhang, Z.; Huang, S.; Bai, X.; Huang, Z.; Zhang, Y.J.; Huang, L.; Tang, W.; Haughn, G.; You, S.; et al. CannabisGDB: A Comprehensive Genomic Database for *Cannabis sativa* L. *Plant Biotechnol. J.* **2021**, *19*, 857–859. [CrossRef]
114. Salha, M.; Adenusi, H.; Dupuis, J.H.; Bodo, E.; Botta, B.; McKenzie, I.; Yada, R.Y.; Farrar, D.H.; Magolan, J.; Tian, K.V.; et al. Bioactivity of the Cannabigerol Cannabinoid and Its Analogues—The Role of 3-Dimensional conformation. *Org. Biomol. Chem.* **2023**, *21*, 4683–4693. [CrossRef]
115. Millan-Linares, M.C.; Rivero-Pino, F.; la Rosa, T.G.; Villanueva, A.; Montserrat-de la Paz, S. Identification, Characterization, and Molecular Docking of Immunomodulatory Oligopeptides from Bioavailable Hempseed Protein Hydrolysates. *Food Res. Int.* **2024**, *176*, 113712. [CrossRef]

116. Nwonuma, C.O.; Atanu, F.O.; Okonkwo, N.C.; Egharevba, G.O.; Udofia, I.A.; Evbuomwan, I.O.; Alejolowo, O.O.; Osemwegie, O.O.; Adelani-Akande, T.; Dogunro, F.A. Evaluation of Anti-Malarial Activity and GC–MS Finger Printing of Cannabis: An In-Vivo and In Silico Approach. *Sci. Afr.* **2022**, *15*, e01108. [CrossRef]
117. Verma, R.P.; Matthews, E.J. Estimation of the Chemical-Induced Eye Injury Using a Weight-of-Evidence (WoE) Battery of 21 Artificial Neural Network (ANN) c-QSAR Models (QSAR-21): Part II: Corrosion Potential. *Regul. Toxicol. Pharmacol.* **2015**, *71*, 331–336. [CrossRef] [PubMed]
118. Toropova, A.P.; Toropov, A.A. (Eds.) *QSPR/QSAR Analysis Using SMILES and Quasi-SMILES*; Challenges and Advances in Computational Chemistry and Physics; Springer International Publishing: Cham, Switzerland, 2023; Volume 33, ISBN 978-3-031-28400-7.
119. Li, S.; Yi, H.; Leng, Q.; Wu, Y.; Mao, Y. New Perspectives on Cancer Clinical Research in the Era of Big Data and Machine Learning. *Surg. Oncol.* **2024**, *52*, 102009. [CrossRef] [PubMed]
120. Sun, K.; Roy, A.; Tobin, J.M. Artificial Intelligence and Machine Learning: Definition of Terms and Current Concepts in Critical Care Research. *J. Crit. Care* **2024**, *82*, 154792. [CrossRef]
121. Zhou, Z.; Wei, L.; Yuan, J.; Cui, J.; Zhang, Z.; Zhuo, W.; Lin, D. Construction Safety Management in the Data-Rich Era: A Hybrid Review Based upon Three Perspectives of Nature of Dataset, Machine Learning Approach, and Research Topic. *Adv. Eng. Inform.* **2023**, *58*, 102144. [CrossRef]
122. Hernán Pérez de la Ossa, D.; Lorente, M.; Gil-Alegre, M.E.; Torres, S.; García-Taboada, E.; Aberturas, M.D.R.; Molpeceres, J.; Velasco, G.; Torres-Suárez, A.I. Local Delivery of Cannabinoid-Loaded Microparticles Inhibits Tumor Growth in a Murine Xenograft Model of Glioblastoma Multiforme. *PLoS ONE* **2013**, *8*, e54795. [CrossRef]
123. Mobaleghol Eslam, H.; Hataminia, F.; Esmaeili, F.; Salami, S.A.; Ghanbari, H.; Amani, A. Preparation of a Nanoemulsion Containing Active Ingredients of Cannabis Extract and Its Application for Glioblastoma: In Vitro and In Vivo Studies. *BMC Pharmacol. Toxicol.* **2024**, *25*, 73. [CrossRef]
124. Reddy, T.S.; Zomer, R.; Mantri, N. Nanoformulations as a Strategy to Overcome the Delivery Limitations of Cannabinoids. *Phytother. Res.* **2023**, *37*, 1526–1538. [CrossRef]
125. Baek, E.J.; Garcia, C.V.; Shin, G.H.; Kim, J.T. Improvement of Thermal and UV-Light Stability of β-Carotene-Loaded Nanoemulsions by Water-Soluble Chitosan Coating. *Int. J. Biol. Macromol.* **2020**, *165*, 1156–1163. [CrossRef] [PubMed]
126. Jin, Y.; Shu, B.; Lou, X.; Wang, K.; Zhai, Y.; Qu, Y.; Song, R.; Liu, F.; Dong, X.; Xu, H. Improvement of Stability and in Vitro Bioaccessibility of Nervonic Acid by Nonionic Surfactant in Protein-Based Nanoemulsions. *Food Biosci.* **2023**, *51*, 102299. [CrossRef]
127. Rave, M.C.; Echeverri, J.D.; Salamanca, C.H. Improvement of the Physical Stability of Oil-in-Water Nanoemulsions Elaborated with Sacha Inchi Oil Employing Ultra-High-Pressure Homogenization. *J. Food Eng.* **2020**, *273*, 109801. [CrossRef]
128. Yao, X.; Teng, W.; Wang, J.; Wang, Y.; Zhang, Y.; Cao, J. Polyglycerol Polyricinoleate and Lecithin Stabilized Water in Oil Nanoemulsions for Sugaring Beijing Roast Duck: Preparation, Stability Mechanisms and Color Improvement. *Food Chem.* **2024**, *447*, 138979. [CrossRef] [PubMed]
129. Ceprián, M.; Jiménez-Sánchez, L.; Vargas, C.; Barata, L.; Hind, W.; Martínez-Orgado, J. Cannabidiol Reduces Brain Damage and Improves Functional Recovery in a Neonatal Rat Model of Arterial Ischemic Stroke. *Neuropharmacology* **2017**, *116*, 151–159. [CrossRef] [PubMed]
130. Victor, T.R.; Hage, Z.; Tsirka, S.E. Prophylactic Administration of Cannabidiol Reduces Microglial Inflammatory Response to Kainate-Induced Seizures and Neurogenesis. *Neuroscience* **2022**, *500*, 1–11. [CrossRef] [PubMed]
131. Vilela, L.R.; Lima, I.V.; Kunsch, É.B.; Pinto, H.P.P.; de Miranda, A.S.; Vieira, É.L.M.; de Oliveira, A.C.P.; Moraes, M.F.D.; Teixeira, A.L.; Moreira, F.A. Anticonvulsant Effect of Cannabidiol in the Pentylenetetrazole Model: Pharmacological Mechanisms, Electroencephalographic Profile, and Brain Cytokine Levels. *Epilepsy Behav.* **2017**, *75*, 29–35. [CrossRef] [PubMed]
132. Yang, M.; Lv, Y.; Tian, X.; Lou, J.; An, R.; Zhang, Q.; Li, M.; Xu, L.; Dong, Z. Neuroprotective Effect of β-Caryophyllene on Cerebral Ischemia-Reperfusion Injury via Regulation of Necroptotic Neuronal Death and Inflammation: In Vivo and In Vitro. *Front. Neurosci.* **2017**, *11*, 583. [CrossRef]
133. Gorelick, D.A.; Heishman, S.J.; Preston, K.L.; Nelson, R.A.; Moolchan, E.T.; Huestis, M.A. The Cannabinoid CB1 Receptor Antagonist Rimonabant Attenuates the Hypotensive Effect of Smoked Marijuana in Male Smokers. *Am. Heart J.* **2006**, *151*, 754.e1–754.e5. [CrossRef]
134. Hodcroft, C.J.; Rossiter, M.C.; Buch, A.N. Cannabis-Associated Myocardial Infarction in a Young Man with Normal Coronary Arteries. *J. Emerg. Med.* **2014**, *47*, 277–281. [CrossRef]
135. Jeffers, A.M.; Glantz, S.; Byers, A.L.; Keyhani, S. Association of Cannabis Use With Cardiovascular Outcomes among US Adults. *J. Am. Heart Assoc.* **2024**, *13*, e030178. [CrossRef]
136. Bashashati, M.; Storr, M.A.; Nikas, S.P.; Wood, J.T.; Godlewski, G.; Liu, J.; Ho, W.; Keenan, C.M.; Zhang, H.; Alapafuja, S.O.; et al. Inhibiting Fatty Acid Amide Hydrolase Normalizes Endotoxin-Induced Enhanced Gastrointestinal Motility in Mice. *Br. J. Pharmacol.* **2012**, *165*, 1556–1571. [CrossRef]
137. Duncan, M.; Mouihate, A.; Mackie, K.; Keenan, C.M.; Buckley, N.E.; Davison, J.S.; Patel, K.D.; Pittman, Q.J.; Sharkey, K.A. Cannabinoid CB2 Receptors in the Enteric Nervous System Modulate Gastrointestinal Contractility in Lipopolysaccharide-Treated Rats. *Am. J. Physiol. Gastrointest. Liver Physiol.* **2008**, *295*, G78. [CrossRef]

138. Jamontt, J.M.; Molleman, A.; Pertwee, R.G.; Parsons, M.E. The Effects of Δ9-Tetrahydrocannabinol and Cannabidiol Alone and in Combination on Damage, Inflammation and In Vitro Motility Disturbances in Rat Colitis. *Br. J. Pharmacol.* **2010**, *160*, 712 [CrossRef] [PubMed]
139. Skinner, C.M.; Nookaew, I.; Ewing, L.E.; Wongsurawat, T.; Jenjaroenpun, P.; Quick, C.M.; Yee, E.U.; Piccolo, B.D.; ElSohly, M.; Walker, L.A.; et al. Potential Probiotic or Trigger of Gut Inflammation—The Janus-Faced Nature of Cannabidiol-Rich Cannabis Extract. *J. Diet. Suppl.* **2020**, *17*, 543–560. [CrossRef] [PubMed]

BioTech

MDPI

Article

Enhanced Bioactive Compounds and Antioxidant Activity in Germinated Seeds of the New Peanut Variety

Hwan-Hee Yu [1], Jong-Suk Park [2] and Sanghyun Lee [1,3,*]

[1] Department of Plant Science and Technology, Chung-Ang University, Anseong 17546, Republic of Korea; hhy120659@cau.ac.kr
[2] Jeonbuk State Agricultural Research & Extension Services, Iksan 54591, Republic of Korea; jspark88@korea.kr
[3] Natural Product Institute of Science and Technology, Anseong 17546, Republic of Korea
* Correspondence: slee@cau.ac.kr; Tel.: +82-31-670-4688; Fax: +82-31-676-4686

Abstract: The sprout market in Korea is expanding as consumers seek healthier food options and farmers strive to increase added value and competitiveness. This study examined the changes in the phytochemical composition of Sinpalkwang (SPK), a peanut variety developed in Korea, during germination. Four samples (SPK1, SPK2, SPK3, and SPK4) were collected at different growth stages and analyzed for total polyphenol content (TPC), total flavonoid content (TFC), and antioxidant activities using $ABTS^+$ and DPPH assays. The levels of *trans*-resveratrol and soyasaponin Bb were quantified using high-performance liquid chromatography (HPLC) with a photo-diode array (PDA). Among the samples, SPK2 exhibited the highest TFC (1.61 mg QE/g ext.) and *trans*-resveratrol content (0.054 mg/g ext.), while SPK4 showed the highest TPC (29.38 mg TAE/g ext.) and soyasaponin Bb content (6.543 mg/g ext.). In terms of radical scavenging activities, SPK2 and SPK3 performed best in the $ABTS^+$ and DPPH assays, respectively. Germinated samples demonstrated strong results across all analyses, highlighting the benefits of germination in enhancing phytochemical properties. This study provides foundational information on the phytochemical composition of SPK and the effects of germination. Future research will focus on optimizing germination conditions to further enhance the functionality and value of this Korean-bred variety as a source of high-value bioactive ingredients.

Keywords: Sinpalkwang; *trans*-resveratrol; soyasaponin Bb; HPLC method; peanut

Key Contribution: The germination of peanut seeds can enhance the content of bioactive compound beneficial to humans.

1. Introduction

Peanuts (*Arachis hypogaea* L.) are a member of the Leguminosae family and are consumed worldwide due to their rich nutrients, including protein, fatty acid, minerals, vitamins, and phytochemicals. Peanuts are cultivated in Asia, Africa, and the Americas, serving as versatile ingredients in cooking, snacks, spreads, butter, and oils [1,2]. Phytochemicals in peanuts, such as carotenoids, stilbene resveratrol, proanthocyanidins, and catechins, have been shown to offer significant health benefits even in small amounts [2]. Numerous studies have demonstrated their efficacy, and some are now commercially produced through artificial synthesis. However, consuming naturally occurring phytochemicals is considered safer due to the reduced risk of side effects or toxicity [3].

Germination is a straightforward and effective method to enhance phytochemical content. During germination, phytochemical accumulation increases, leading to improved physiological activity [4,5]. Studies on legume germination also indicate that it reduces antinutrient compounds, which inhibit nutrient absorption [6]. Germination is eco-friendly, requiring only clean water, no pesticides, and a short growth period [7].

In Korea, the growing interest in health has popularized the consumption of various sprout seeds, including kohlrabi, red radish, broccoli, cabbage, radish sprouts, peas, lentils, and mung beans, as food ingredients. These sprouts also hold high commercial value as natural functional ingredients and highly bioactive food materials. Peanut sprouts, primarily processed into powder form, are used in products like rice cakes and baked goods. A study on madeleines containing peanut sprout powder revealed that increasing the amount of peanut sprout powder resulted in higher total polyphenol content, total flavonoid content, and antioxidant activity [8].

Sinpalkwang (SPK) is a high-yield peanut variety developed through breeding in Korea. It was developed via the cross-fertilization of the large-grain Palkwang and the short-stemmed PI156649 varieties by the National Institute of Crop Science, Jeonbuk, Republic of Korea [9]. SPK is characterized by large grains and short stems [9]. SPK has lower crude fat and higher linoleic acid compared to Daekwang, which was previously cultivated in Korea. Selected as a new variety in 2012, SPK continues to be widely cultivated in Korea [9]. Despite ongoing efforts to develop new varieties through breeding, research on peanut varieties grown in Korea remains limited. In the case of SPK, studies on its phytochemical composition are scarce, primarily focusing on *trans*-resveratrol and soyasaponin Bb.

trans-Resveratrol, a polyphenol found in grape skins, red or purple fruits, peanuts, and wine, has gained significant attention as a potential factor in the "French paradox", where the French exhibit a low incidence of cardiovascular disease despite a diet high in saturated fats [10,11]. This compound not only possesses strong antioxidant properties but also promotes cardiovascular health [10]. However, the oral bioavailability of *trans*-resveratrol is limited; thus, recent studies have explored the use of nanoparticles, such as solid lipid nanoparticles, to enhance its delivery to target organs [12,13]. In addition, *trans*-resveratrol has been investigated as an activator of Sirtuin 1 (SIRT1), which regulates the post-translational modifications of proteins, contributing to the longevity of cells [14,15]. At NAD^+ concentrations above 100 μM, *trans*-resveratrol can increase SIRT1 activity eightfold by reducing the K_M of the deacetylation reaction [16,17]. SIRT1 activation through either resveratrol or calorie restriction consequently improves hepatic insulin sensitivity and regulates cholesterol and fatty acid metabolism in the liver [18].

Soyasaponin is a component that exists in many legume crops regardless of species [19]. Soyasaponins, which consist of aglycone and oligosaccharide groups, can be categorized into A, B, and E types based on their aglycone structure [20]. Soyasaponin A contains soyasapogenol A as its aglycone. Similarly, soyasaponin B and E have soyasapogenol B and E as their respective aglycones [21]. Soyasaponin Bb is the main monomeric compound of soyasaponins [22]. Their type and content vary between seed species, growth conditions, and growth stage [23]. Processing methods also impact soyasaponin levels, which are lower during boiling and higher when soaking or sprouting [23]. In particular, soybean sprouts have been shown to raise soyasaponin levels 3.2-fold compared to mature soybeans because the plant synthesizes soyasaponins as a defense mechanism during germination [24]. Higher soyasaponin levels are associated with stronger antioxidant efficacy, and at higher concentrations, they exert an inhibitory effect on the growth and survival of colon cancer cells [25]. Additionally, soyasaponins help prevent fat accumulation in the liver and reduce adipose tissue by promoting fecal excretion [26].

The present study aimed to evaluate the content of bioactive compounds in SPK using HPLC-PDA and their relationship with antioxidant efficacy, with a focus on *trans*-resveratrol and soyasaponins. Since SPK is a high-yielding variety, it will have high nutritional value if the time when bioactive compounds are most active is identified. This also yield beneficial potential for humans, as it has a higher antioxidant effect induced by enhanced bioactive compounds.

2. Materials and Methods

2.1. Plant Materials

As shown in Figure 1, SPK was classified into four serial growth stages. SPK 1 is the dried SPK peanut. SPK 2 is the stage where the sprouts are not visible outside the cotyledons, while SPK 3 has sprouts that are visible about 1 cm outside the cotyledons. In SPK 4, the sprouts are visible about 3 cm outside the cotyledons. All samples were provided by Jeonbuk State Agricultural Research, Iksan, Republic of Korea. Voucher specimens were stored in Jeonbuk State Agricultural Research, Iksan, Republic of Korea. After collecting, the samples (voucher No. LEE2024-0428) were analyzed in powdered form in July 2024.

Figure 1. SPK samples by sampling time: (**A**) SPK 1, (**B**) SPK 2, (**C**) SPK 3, and (**D**) SPK 4.

2.2. Growth Conditions

Peanut seeds of the SPK variety were soaked in water for 24 h on 28 April 2024. Subsequently, the seeds were transferred to a cultivation tray and subjected to aeroponic cultivation. The spray system was programmed to spray water for 30 s at 10 min intervals. The aeroponics procedure was conducted entirely in darkness and continued until 10 May 2024.

2.3. Instruments and Reagents

trans-Resveratrol and soyasaponins were analyzed using HPLC (Waters Alliance system e2695 Separations Module, Milford, MA, USA) with a 2998 PDA and an INNO C18 column (4.6 × 250 mm, 5 μm). HPLC-grade water (H_2O) and methanol (MeOH) were purchased from Honeywell (Burdick and Jackson, Muskegon, MI, USA). Ethanol (EtOH) was purchased from Samchun Chemicals (Pyeongtaek, Republic of Korea). Acetonitrile (ACN; HPLC grade) was obtained from J. T. Baker (Phillipsburg, PA, USA). Acetic acid and phosphoric acid (HPLC grade) were purchased from Fisher Scientific (Loughborough, Leicestershire, UK). Polyvinylidene fluoride (PVDF) membrane filter was obtained from Hyundai Micro (Seoul, Republic of Korea). In addition, *trans*-resveratrol (**1**), soyasaponin

Aa (**2**), soyasaponin Ab (**3**), soyasaponin Ba (**4**), and soyasaponin Bb (**5**) were supplied by Natural Product Institute of Science and Technology (www.nist.re.kr; accessed on 5 July 2024), Anseong, Republic of Korea (Figures 2 and 3).

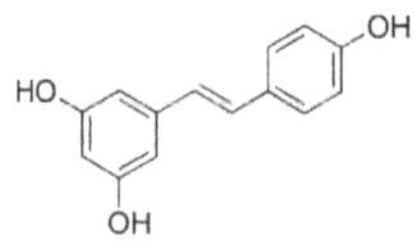

Figure 2. Chemical structure of *trans*-resveratrol (**1**).

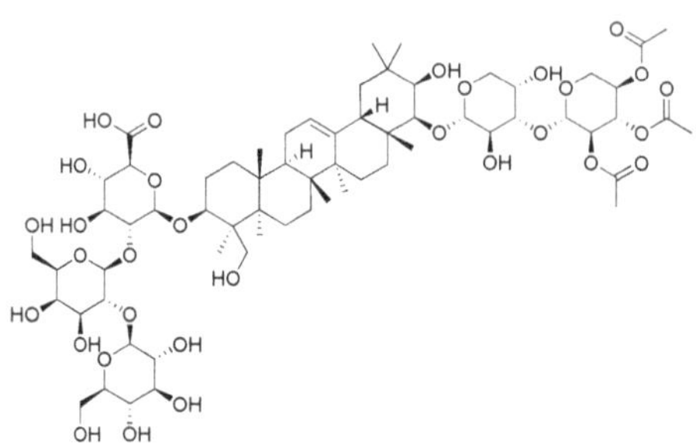

2

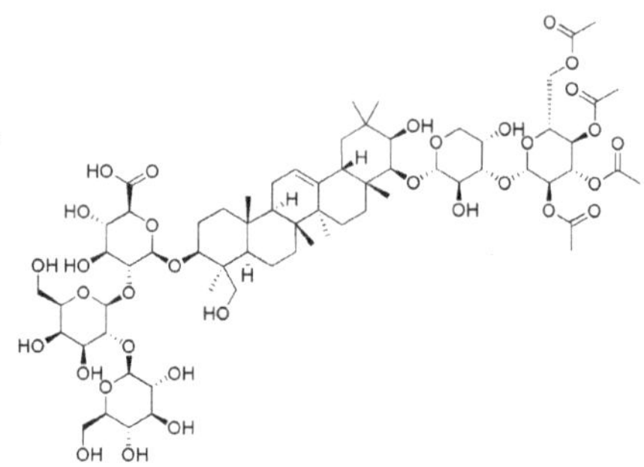

3

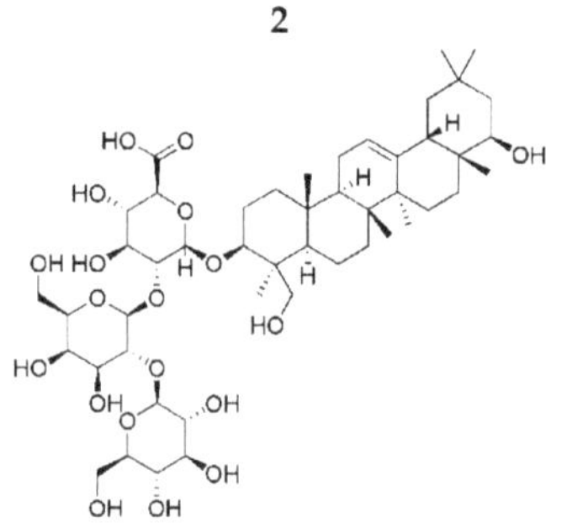

4

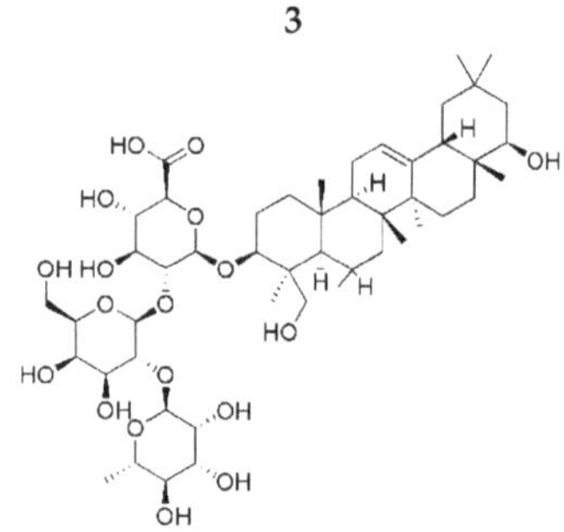

5

Figure 3. Chemical structures of soyasaponins Aa (**2**), Ab (**3**), Ba (**4**), and Bb (**5**).

2.4. Preparation of SPK Extracts

Dried SPK samples were ground into a powder, mixed with 240 mL of 95.0% EtOH, and heated in a reflux condenser for 3 h. The ratio of dry sample to solvent was set at 1:40. The solution extracted from the sample was poured onto filter paper to remove the solid matter. The filtered solution was then concentrated using a rotary evaporator. The dried solid-state extracts were used to prepare samples for subsequent experiments.

2.5. Sample Preparation for Antioxidant Activities

The dried extracts of SPK 1, SPK 2, SPK 3, and SPK 4 (50 mg each) were individually sonicated together with 50% MeOH for 40 min until they were completely dissolved. The solutions were centrifuged for 10 min, and the supernatant was collected and filtered through a PVDF membrane filter with a pore size of 0.45 μm. Antioxidant efficacy was assessed based on IC_{50}, which is the concentration required to reduce 2,2′-azino-bis-(3-ethylbenzothiazoline-6-sulfonic) acid ($ABTS^+$) and 2,2-diphenyl-1-picrylhydrazyl (DPPH) radicals by 50% [27].

2.6. ABTS+ Radical Scavenging Activity of SPK

$ABTS^+$ radical scavenging activity was evaluated using a 96-well plate (SPL Life Sciences, Pocheon, Republic of Korea). $ABTS^+$ was acquired from Thermo Fisher Scientific Inc. (Waltham, MA, USA). The dried SPK extracts were mixed with 50% MeOH. The $ABTS^+$ and ascorbic acid stock solutions (A0278-100G, Sigma-Aldrich, St. Louis, MO, USA) were combined with water and used as the working solutions. The ascorbic acid working solution (10 μL) and the SPK samples (10 μL) were combined in each well, and 200 μL of the $ABTS^+$ working solution was added. The reaction was left to occur for 30 min in the dark. The absorbance of the $ABTS^+$ solution was measured at 734 nm using a microplate reader (BioTek Epoch, Santa Clara, CA, USA).

2.7. DPPH Radical Scavenging Activity of SPK

DPPH radical scavenging activity was performed using a 96-well microplate. DPPH was acquired from Thermo Fisher Scientific Inc. The dried SPK extracts were mixed with 50% MeOH. A stock solution of ascorbic acid was mixed with water and used as a working solution. A DPPH working solution was produced by mixing the DPPH stock solution with 95% MeOH. The ascorbic acid working solution (10 μL) and the individual samples (10 μL) were combined in the wells of the microplate, to which 200 μL of DPPH working solution was subsequently added. The reaction was left to proceed in the dark for 30 min. The absorbance of the DPPH reaction at 514 nm was then measured using a microplate reader.

2.8. Total Polyphenol Content (TPC) of SPK

The TPC of the SPK samples was measured using a 96-well plate. Folin and Ciocalteu's phenol reagent (2N) and tannic acid (ACS reagent) were purchased from Sigma-Aldrich (Saint Louis, MA, USA). Sodium carbonate anhydrous was acquired from Daejung Chemicals and Metals (Siheung, Republic of Korea). The tannic acid was utilized as a standard solution in TPC analysis. Therefore, tannic acid of 5000 ppm was added to water in six concentrations to draw a calibration curve. All of the samples were completely liquefied by sonication and then purified using a 0.45 μm membrane filter. The 7.5% sodium carbonate solution was employed as the interaction solution. After taking 60 μL of each SPK sample, 40 μL of 2N Folin and Ciocalteu's phenol reagent, and 100 μL of the 7.5% sodium carbonate solution together, the optical density was measured at 760 nm using a microplate reader.

2.9. Total Flavonoid Content (TFC) of SPK

The TFC of the SPK samples was measured using a 96-well plate. EtOH, water, 5000 ppm quercetin (Sigma-Aldrich, MA, USA), and 2% aluminum (III) chloride hexahydrate were used for TFC evaluation. The 2% aluminum (III) carbohydrate solution was produced by mixing aluminum chloride hexahydrate (Sigma-Aldrich, MA, USA) and the water. A calibration curve was produced in this way for the TPC using a quercetin solution instead of tannic acid. After adding 100 μL of the 2% aluminum (III) carbohydrate solution to 100 μL of each sample, the mixture was left in the dark for 10 min. The optical density was then measured at 430 nm using a microplate reader.

2.10. HPLC Analysis of Trans-Resveratrol

HPLC analysis of *trans*-resveratrol was conducted using a reverse phase column (INNO C_{18} column, 4.6 × 250 nm, Youngjin Biochrom, Seongnam, Republic of Korea) at 35 °C. The standard solution was prepared by mixing the *trans*-resveratrol (**1**) with 80% MeOH to produce a concentration of 1000 ppm (1 mg/mL) and sonicating it for 20 min. The dried SPK extracts were dissolved in 80% MeOH under sonication for 20 min to produce a 50,000 ppm (50 mg/mL) solution. The resulting solution was filtered and diluted based on the HPLC results recorded at 310 nm. All of the solutions were injected at 5 μL each maintained at 35 °C. Mobile phases A and B were water containing 0.5% acetic acid and ACN, respectively, with a flow rate of 1.0 mL/min. The A/B ratio was 85:15 for 5 min, followed by 77:23 for 10 min, before following a gentle gradient to finish at 70:30. After this process, the column was washed with 100% ACN for 15 min. After washing, the A/B ratio was set at 85:15 for 10 min for stabilization.

2.11. HPLC Analysis of Soyasaponins Aa, Ab, Ba, and Bb

Soyasaponins Aa, Ab, Ba, and Bb were analyzed using HPLC-PDA with a reverse-phase column (YMC-Pack Pro C18 column, 4.6 × 250 nm, 5 μm, YMC Korea, Seongnam, Republic of Korea) at 35 °C. Standard solutions of soyasaponins Aa, Ab, Ba, and Bb were prepared by combining 1 mg of each with 1 mL of 80% MeOH (1000 ppm). To produce four peaks in one chromatogram, identical amounts of the four standard solutions were mixed to produce a single 250 ppm solution. Each SPK extract was dissolved in 80% MeOH and sonicated for 20 min to produce a 10,000 ppm (10 mg/mL) solution. All of the solutions used in the soyasaponin tests were injected at 20 μL at 35 °C. Mobile phases A and B consisted of water containing 0.1% phosphoric acid and ACN, respectively, with a flow rate of 1.0 mL/min. The A/B ratio was set at 90:10 for 10 min, 70:30 for 20 min, 60:40 for 15 min, and finally 45:55 for 5 min. For washing and stabilization, mobile phases A and B were used at 0:100 for 8 min and 90:10 for 15 min.

2.12. Calibration Curves

To determine the quantity of *trans*-resveratrol in peanuts and sprouts, a calibration curve for *trans*-resveratrol was produced using seven standard solutions with different concentrations (0.98, 1.95, 3.91, 7.81, 15.62, 31.25, and 62.5 ppm). The linear equation for the calibration curve was $y = 36262x - 6215.9$, with strong linearity observed ($r^2 > 0.9999$). To produce the calibration curve for soyasaponin Bb, six solutions with concentrations of 3.91, 7.81, 15.62, 31.25, 62.5, and 125 ppm were used. The linear equation was $y = 2618.3x + 11129$, with a linear regression coefficient (r^2) of 0.9978.

2.13. Statistical Analysis

All experiments were run three times using the same method to ensure the reproducibility and accuracy of the results. The experimental results were expressed as the sum and difference of the mean and standard deviation. Statistical analysis was conducted using Minitab (Version 16, Minitab, Inc., State College, PA, USA). Significant differences between samples were determined using ANOVA and Tukey's tests, with a significance level of $p < 0.05$.

3. Results and Discussion

3.1. SPK Extracts

The dry weight, extract weight, and yield of the SPK samples are summarized in Table 1. The highest yield was achieved for SPK 3 (55.0%), which produced an extract with a mixture of solid and oil. The yield of peanut extracts can depend on the solvent used for extraction (e.g., MeOH, EtOH, and water). Yu et al. reported that roasted peanut peels had the highest phenol content when extracted with 80% EtOH [28]. Extraction using EtOH has a high extraction yield, particularly when extracting bioactive compounds [29]. Therefore, in this study, high-purity 95% EtOH was used. Given that the lowest yield was 50.0% for SPK 1, it was shown that an appropriate solvent was used.

Table 1. Extract yield of SPK samples.

Sample	DW (g)	Ext. Weight (g)	Yield (%)
SPK 1	6	3.0	50.0
SPK 2	6	3.1	51.7
SPK 3	6	3.3	55.0
SPK 4	6	3.2	53.3

Note: DW, dry weight; Ext., extract.

3.2. *ABTS⁺ and DPPH Radical Scavenging Activity of SPK*

Spectrophotometric $ABTS^+$ and DPPH assays are widely used to evaluate redox activity based on the scavenging of free radicals and are useful indicators of antioxidant ability [30]. The $ABTS^+$ and DPPH radical scavenging activity of SPK is presented in Table 2. In $ABTS^+$ assays, ABTS radicals gain electrons when reduced by antioxidants, causing the blue or green $ABTS^+$ to become transparent [30]. SPK 2 (6.20 mg/mL) had the highest antioxidant effect in the $ABTS^+$ assays, while SPK 1 (23.32 mg/mL) exhibited the lowest antioxidant activity. When radical scavenging activity was evaluated using DPPH, SPK 3 (14.48 mg/mL) was found to have the strongest effect, while the highest IC_{50} was produced by SPK 1 (44.24 mg/mL). Thus, the antioxidant effect was strongest when the sprouts first emerged from the cotyledons, but it was almost nonexistent in the ungerminated seeds.

Table 2. $ABTS^+$-DPPH radical scavenging activity of SPK extracts.

Sample	IC_{50} (mg/mL)	
	$ABTS^+$	DPPH
SPK 1	23.32 ± 0.30 [a]	44.24 ± 0.66 [a]
SPK 2	6.20 ± 0.08 [d]	17.61 ± 0.80 [b]
SPK 3	11.75 ± 0.02 [b]	14.48 ± 0.87 [c]
SPK 4	7.63 ± 0.08 [c]	15.05 ± 0.71 [c]
Ascorbic acid	0.11 ± 0.00	0.13 ± 0.00

Note: [a–d] different letters in the same column indicate significant statistical differences ($p < 0.05$).

When comparing the results of the $ABTS^+$ and DPPH assays, the $ABTS^+$ assays demonstrated a higher antioxidant effect for the same samples. Floegel et al. reported that, when comparing the oxygen radical absorbance of phenolic and flavonoid compounds from 50 fruits and vegetables, the results of the ABTS+ assay showed a higher correlation with the antioxidant effect than the results of the DPPH assay [30]. Zhou et al. also reported that, when comparing the antioxidant capacity of peanut sprouts grown in different regions, $ABTS^+$ assays returned higher values than DPPH assays, likely due to the relatively low reactivity of DPPH [31].

3.3. *TPC and TFC of SPK*

The TPC and TFC of the SPK extracts are presented in Table 3. Thaipong et al. reported that the three groups (e.g., vitamins, phenolics, and carotenoids) influenced the effectiveness of antioxidants [32].

The sample with the highest polyphenol content was SPK 4 (29.38 mg TAE/g ext.). In contrast, SPK 1 (16.92 mg TAE/g ext.) had the lowest polyphenol content. Considering the statistical differences, it reaffirms that the samples with good antioxidant effects had high TPC, and samples with poor antioxidant effects had low TPC. [33]. The sample with the highest flavonoid content was SPK 2 (1.61 mg QE/g ext.), which was derived from peanuts with freshly sprouted cotyledons. The sample with the lowest flavonoid content was SPK 3 (0.84 mg QE/g ext.). Overall, however, the TFC levels of the four samples were not statistically different. Peanuts contain various flavonoids, such as anthocyanins and catechins, that are powerful antioxidants.

Table 3. TPC and TFC of SPK extracts.

Sample	TPC (mg TAE/g ext.)	TFC (mg QE/g ext.)
SPK 1	16.92 ± 1.93 [c]	0.86 ± 0.69 [a]
SPK 2	28.22 ± 1.11 [a]	1.61 ± 1.03 [a]
SPK 3	23.16 ± 0.04 [b]	0.84 ± 0.34 [a]
SPK 4	29.38 ± 2.18 [a]	0.91 ± 0.32 [a]

Note: [a–c] different letters in the same column indicate significant statistical differences ($p < 0.05$). TAE, tannic acid equivalent; QE, quercetin equivalent.

However, because antioxidant effects are influenced by both flavonoids and non-flavonoids, TFC alone cannot accurately reflect antioxidant activity. In peanuts, in the early stages of germination, the synthesis of new flavonoids is triggered, increasing TFC. However, as germination continues, water is retained, leading to the loss of water-soluble flavonoids such as anthocyanins and catechins [31]. When sprouts reach about 1 cm in length, the TFC content decreases considerably. In addition, SPK 3 and SPK 4 are exposed to more light and oxygen as the shoot emerges from the cotyledon, and flavonoids are broken down by light and oxidized when exposed to oxygen [31]. In addition, Tran et al. reported that the content of flavonoids can vary depending on how a sample is extracted [34]. In particular, they found that EtOH was more efficient than MeOH in measuring the TFC of seeds and sprouts. However, even when EtOH was used as the extraction solvent, they reported that the TFC results still varied depending on the extraction method. In particular, the amount of TFC extracted using a reflux condenser was significantly lower than that using ultrasonic extraction. This is because reflux extraction requires heat energy to vaporize the solvent, and flavonoids can be degraded by heat. The TFC content can also vary depending on the extraction time, with longer exposure to heat leading to more flavonoids being broken down [32]. It is thus possible that, in the present study, the TFC content was reduced due to the continuous heat applied during extraction and the action of light and oxidation.

3.4. Content of Trans-Resveratrol

MS and NMR data were not available, as *trans*-resveratrol was not isolated from the sample. Instead, the presence of *trans*-resveratrol in peanuts is supported by the previously published literature [2,10,11,18]. Within the UV spectrum (210–400 nm), *trans*-resveratrol has a high absorbance at 310 nm. Based on this, HPLC was performed at a wavelength of 310 nm to determine the absorbance of a 50,000 ppm *trans*-resveratrol standard. A shoulder-less and symmetrical peak was subsequently found at 24.80 min. When screening the chromatograms of the SPK samples, a single peak at 24.80 min was observed (Figure 4). Based on the area of the peak, the sample with the highest *trans*-resveratrol content was SPK 2 (0.054 mg/g ext.). In contrast, only trace amounts of *trans*-resveratrol were detected in SPK 1 (Table 4). To obtain additional data supporting the presence of *trans*-resveratrol, UV spectrums of the standard and samples were compared, and it was confirmed that the UV absorption wavelengths of *trans*-resveratrol and the samples of SPK were comparable (Figure 5).

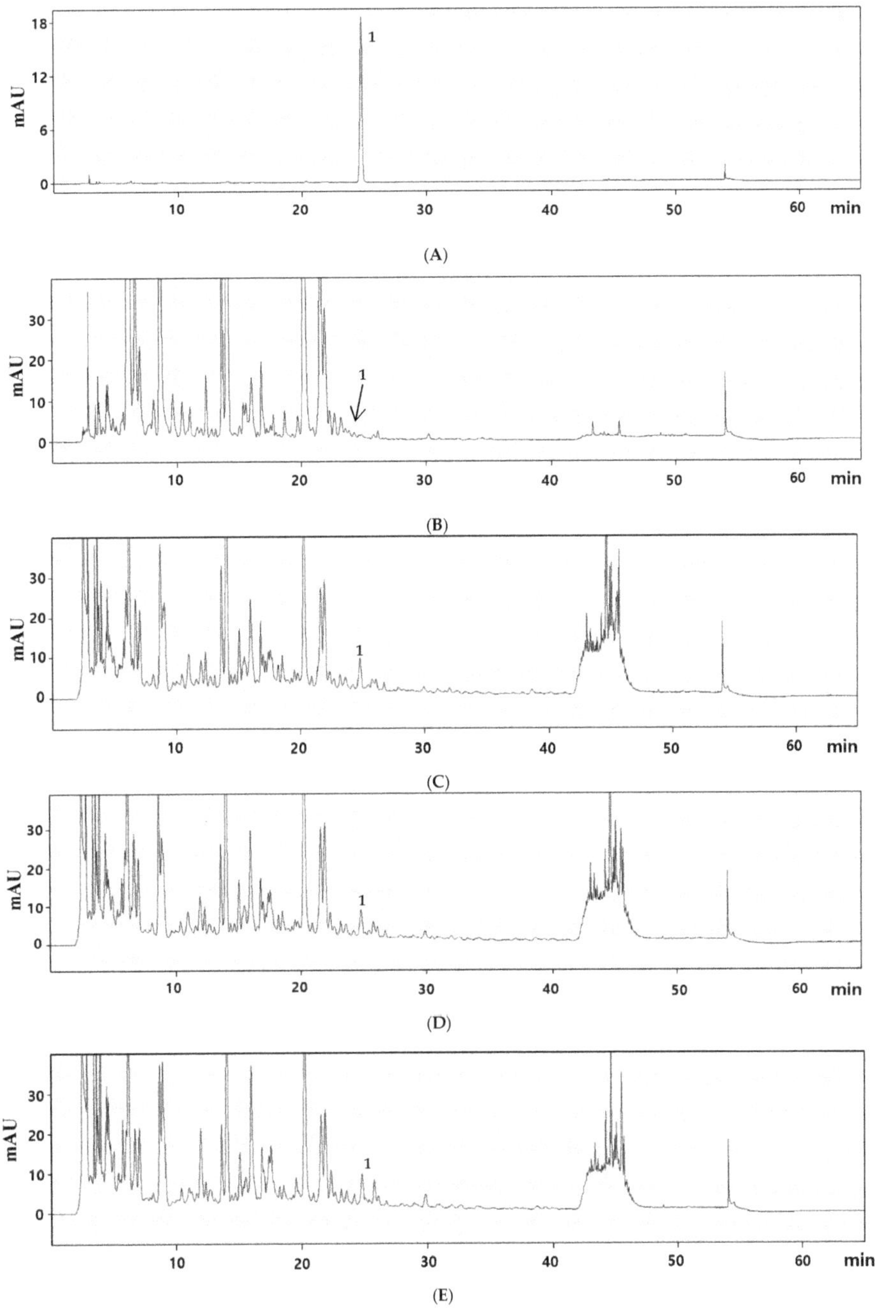

Figure 4. HPLC chromatograms of (**A**) *trans*-resveratrol (**1**), (**B**) SPK 1, (**C**) SPK 2, (**D**) SPK 3, and (**E**) SPK 4.

Table 4. Content of *trans*-resveratrol and soyasaponins of SPK extracts.

Sample	Content (mg/g ext.)					
	1	2	3	4	5	Total
SPK 1	Trace	ND	ND	ND	1.025 ± 0.102 [d]	1.025
SPK 2	0.054 ± 0.001 [a]	ND	ND	ND	2.587 ± 0.102 [c]	2.641
SPK 3	0.045 ± 0.000 [c]	ND	ND	ND	4.149 ± 0.045 [b]	4.194
SPK 4	0.048 ± 0.001 [b]	ND	ND	ND	6.543 ± 0.146 [a]	6.591

Note: [a–d] different letters in the same column indicate significant statistical differences ($p < 0.05$). ext., extract. **1**: *trans*-resveratrol, **2**: soyasaponin Aa, **3**: soyasaponin Ab, **4**: soyasaponin Ba, and **5**: soyasaponin Bb.

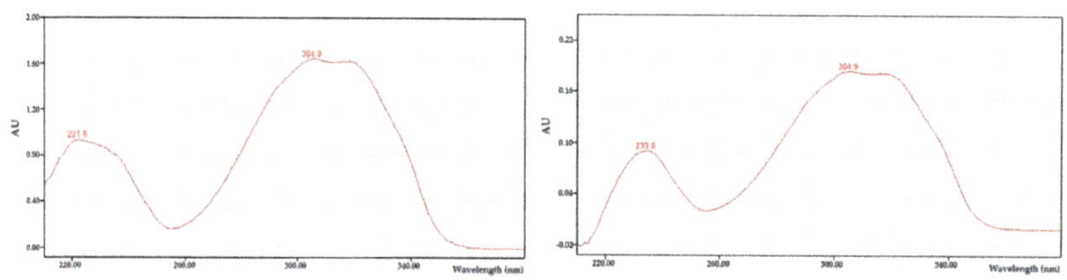

(A) (B)

Figure 5. UV spectrums of (**A**) *trans*-resveratrol and (**B**) SPK 2.

In another study, the *trans*-resveratrol content of raw seeds of Palkwang, one of the parent varieties of SPK, was found to be 0.14 μg [35]. Although the exact values cannot be compared in this study, the *trans*-resveratrol content in SPK after sprouting increased more than 100-fold compared to that in raw seeds of Palkwang. Pae et al. found that the average *trans*-resveratrol content after germination of 37 peanut cultivars in Korea was 43.9 μg, and the *trans*-resveratrol content of Palkwang after germination was 45.8 μg. [36]. The content of *trans*-resveratrol in SPK after germination was also detected similarly.

We hypothesized that, as the shoots grow, more *trans*-resveratrol is generated. There are two possibilities as to why the content of *trans*-resveratrol decreases as peanut sprouts grow: either it is decomposed, or it is used. The structure of resveratrol changes when it reacts with light, and this is more apparent under more intense UV light [37]. Specifically, when *trans*-resveratrol is exposed to UV-B radiation, most of it transforms into other forms within 4 h. However, under laboratory conditions, which primarily involve visible light, *trans*-resveratrol remains stable [38]. Since the seeds were germinated in darkness, and the experiments were conducted indoors, light exposure likely had a minimal effect on *trans*-resveratrol content. Additionally, one study showed that *trans*-resveratrol influences the growth of lettuce aerial parts [39], suggesting that *trans*-resveratrol may have been utilized during the growth of peanut sprouts.

3.5. Content of Soyasponins Aa, Ab, Ba, and Bb

MS and NMR data were not available, as soyasaponin Bb was not isolated from the sample. Instead, the presence of soyasaponin Bb in peanuts is supported by the previously published literature [40,41]. Within the UV spectrum (190–400 nm), soyasaponin group A and B show high absorbance near 210 nm. Because of the low absorption wavelength below 210 nm, UV spectral data were not sufficient to confirm the presence of soyasaponin Bb. Therefore, a spike test was conducted to observe the change in the HPLC chromatogram by adding soyasaponin Bb standard to the SPK sample. When the SPK samples and soyasaponin Bb standard were mixed, the peak area increased, and no new peaks were discovered. Additionally, to confirm the presence of other types of soyasaponin, four standards of soyasaponin, Aa, Ab, Ba, and Bb, were mixed and analyzed by HPLC. The peak of soyasaponin Bb with a retention time of 37.9 min was observed in all SPK samples. However, the peaks for soyasaponin Aa, Ab, and Ba with retention times of 29.5, 31.3, and 36.5 min were not detected in all SPK samples (Figure 6).

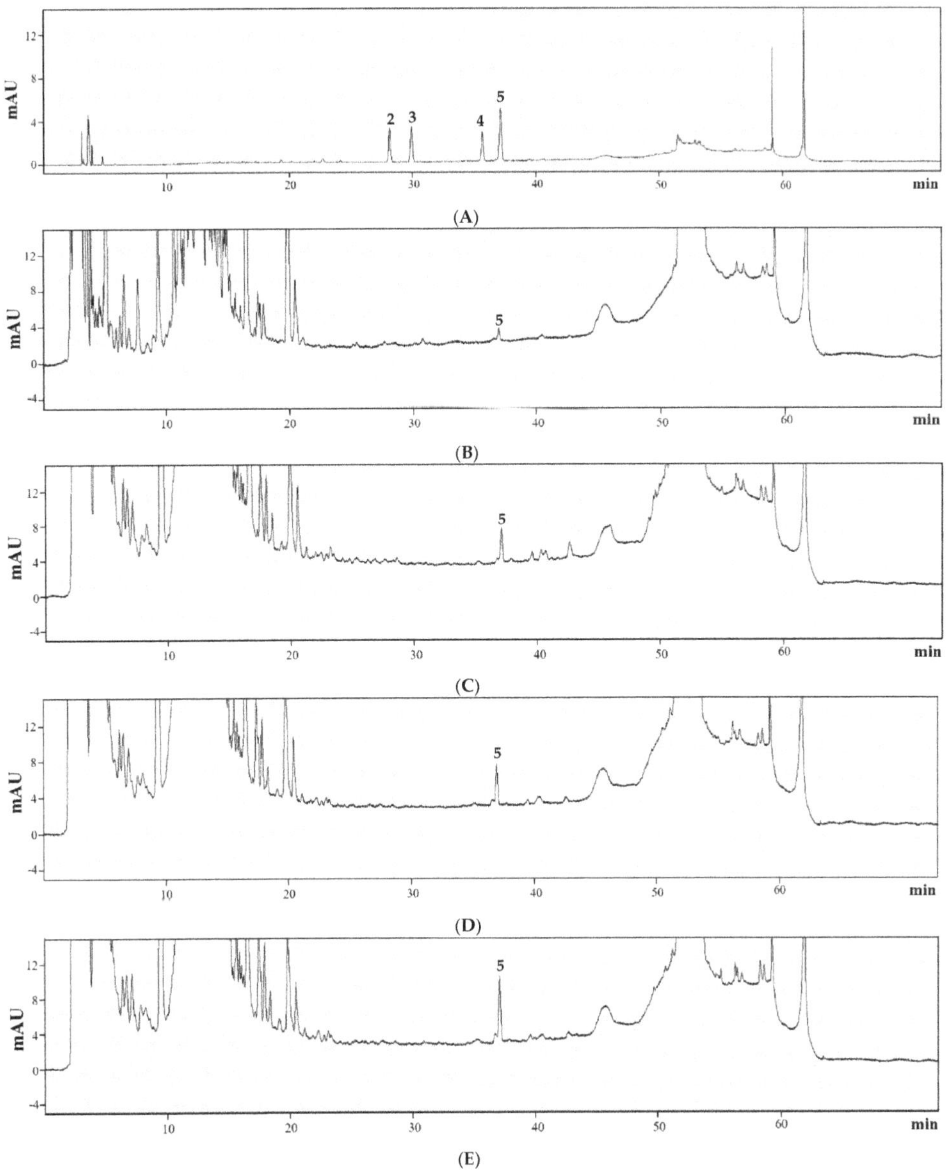

Figure 6. HPLC chromatograms for (**A**) soyasaponins Aa (**2**), Ab (**3**), Ba (**4**), Bb (**5**), (**B**) SPK 1, (**C**) SPK 2, (**D**) SPK 3, and (**E**) SPK 4.

The content of soyasaponin Bb was found to be highest in the SPK 4 extract, which was taken from the longest shoots, but was the lowest in the SPK 1 extract, with a 6.5-fold difference between the two. Similarly, in a study analyzing germinated black beans, the content of soyasaponin Bb was found to increase by approximately 40 mg/100 g after germination [42]. In another study, the content of soyasaponin Bb in soybeans before and after germination was compared, with a 15-fold difference observed between the two [43]. These studies and the results of the present study suggest that soyasaponin Bb is continuously synthesized and accumulates during the germination process [43].

Soyasaponin Bb is a photo-response compound whose content in soybeans changes significantly depending on the presence of light [44]. For example, the soyasaponin Bb content is higher when soybean seeds germinate under light than in the dark [44]. The findings from the experiments conducted in this study are summarized in the following graph (Figure 7).

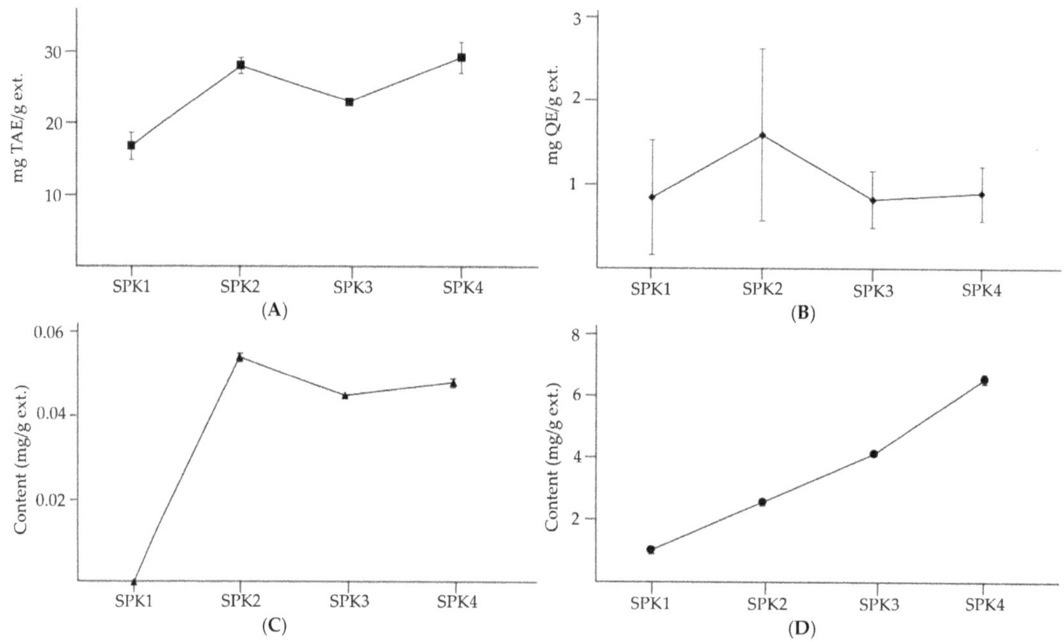

Figure 7. Tendency of SPK for (**A**) TPC, (**B**) TFC, (**C**) content of *trans*-resveratrol, and (**D**) soyasaponin Bb.

4. Conclusions

This study showed that germination had a positive effect on increasing the nutritional value of seeds. The germinated seeds showed high antioxidant effects and increased phenolic compounds and phytochemical nutrients. Our study confirmed that germination, a low-cost and practical method, is very effective in improving nutritional value. SPK2 exhibited the highest TFC (1.61 mg QE/g ext.) and *trans*-resveratrol content (0.054 mg/g ext.), while SPK4 showed the highest TPC (29.38 mg TAE/g ext.) and soyasaponin Bb content (6.543 mg/g ext.). In terms of radical scavenging activities, SPK2 and SPK3 performed best in the $ABTS^+$ and DPPH assays, respectively. Germinated samples demonstrated strong results across all analyses, highlighting the benefits of germination in enhancing phytochemical properties. The increased bioactive compounds through germination and the high production of newly bred SPK have the potential for wide application in nutritional and therapeutic aspects. Further research can be conducted to determine appropriate germination conditions that induce the desired nutritional compositions.

Author Contributions: HPLC/PDA, TPC, TFC analysis, and antioxidant activity, H.-H.Y.; resources and experimental design, J.-S.P.; supervision and writing—review and editing, S.L. All authors have read and agreed to the published version of the manuscript.

Funding: This work was supported by the Regional Innovation Cluster Development (R&D) research projects (P0025455) funded by the Ministry of Trade, Industry, and Energy (MOTIE), Republic of Korea.

Institutional Review Board Statement: Not applicable.

Informed Consent Statement: Not applicable.

Data Availability Statement: The data presented in this study are available on request from the corresponding author.

Acknowledgments: We express our gratitude to Jeonbuk State Agricultural Research and Extension Services (Iksan, Republic of Korea) for generously gifting the samples.

Conflicts of Interest: The authors declare no conflicts of interest.

References

1. Jung, M.; Kim, J.; Ahn, S.M. Factors associated with frequency of peanut consumption in Korea: A national population-based study. *Nutrients* **2020**, *12*, 1207. [CrossRef] [PubMed]
2. King, J.C.; Blumberg, J.; Ingwersen, L.; Jenab, M.; Tucker, K.L. Tree nuts and peanuts as components of a healthy diet. *J. Nutr.* **2008**, *138*, 1736S–1740S. [CrossRef]
3. Kim, J.; Choi, C.H.; Lee, A.Y.; Lee, S. Antioxidant, anticancer, and neuroprotective activities and phytochemical analysis of germinated shoots. *J. Food Biochem.* **2023**, *2023*, 1–9. [CrossRef]
4. Lu, Y.; Wang, F.; Luo, H.; He, W.; Li, D.; Bao, Y.; Zhang, Z.; Zhou, C. Changes in phytochemical profiles, relevant enzyme activity and antioxidant capacity of different germinated maize varieties. *Food Biosci.* **2023**, *56*, 103410. [CrossRef]
5. Lee, Y.-K.; Jung, Y.-S.; Cho, E.-C.; Kim, D.-H.; Shin, K.-O. Physiological Activity and Nutritional Properties of Peanut Sprout Extracts. *J. Korean Soc. Food Sci. Nutr.* **2024**, *53*, 599–607. [CrossRef]
6. Gonçalves, J.P.; Gasparini, K.; Picoli, E.A.T.; Costa, M.D.L.; Araujo, W.L.; Zsögön, A.; Ribeiro, D.M. Metabolic control of seed germination in legumes. *J. Plant Physiol.* **2024**, *295*, 154206. [CrossRef] [PubMed]
7. Shin, G.-H.; Lee, Y.-J.; Kim, J.-H.; Kim, Y.-H.; Kim, D.-B.; Lee, J.S.; Lim, J.-H.; Lee, O.-H. Antioxidant activities of commonly used Brassica spp. sprout vegetables in Korea. *J. Korean Soc. Food Preserv.* **2014**, *21*, 587–592. [CrossRef]
8. Lee, Y.-K.; Jung, Y.-S.; Cho, E.-C.; Kim, D.-H.; Shin, K.-O. Quality Characteristics of Madeleines Prepared with the Addition of Peanut Sprout Powder. *J. East Asian Soc. Diet. Life* **2024**, *34*, 298–309. [CrossRef]
9. Pae, S.B.; Hwang, C.D.; Kim, S.U.; Lee, M.-H.; Shim, K.B.; Park, C.; Lee, C.-K.; Baek, I.Y.; Lee, J.-K. A new large grain and high-yielding Virginia type peanut cultivar 'Sinpalkwang'. *Korean J. Breed. Sci.* **2016**, *48*, 66–71. [CrossRef]
10. Das, M.; Das, D.K. Resveratrol and cardiovascular health. *Mol. Asp. Med.* **2010**, *31*, 503–512. [CrossRef]
11. Uggioni, M.L.R.; Ronsani, L.; Motta, S.; Júnior, J.C.D.; Marçal, F.; Ferraz, S.D.; Rosa, M.I.D.; Colonetti, T. Effects of resveratrol on the lipid profile of post-menopause women: Systematic review and meta-analysis. *Nutr. Metab. Cardiovasc. Dis.* **2024**, 103827. [CrossRef] [PubMed]
12. Lindner, G.R.; Khalil, N.M.; Mainardes, R.M. Resveratrol-loaded polymeric nanoparticles: Validation of an HPLC-PDA method to determine the drug entrapment and evaluation of its antioxidant activity. *Sci. World J.* **2013**. [CrossRef]
13. Rigon, R.B.; Fachinetti, N.; Severino, P.; Durazzo, A.; Lucarini, M.; Atanasov, A.G.; El Mamouni, S.; Chorilli, M.; Santini, A.; Souto, E.B. Quantification of *trans*-resveratrol-loaded solid lipid nanoparticles by a validated reverse-phase HPLC photodiode array. *Appl. Sci.* **2019**, *9*, 4961. [CrossRef]
14. Langley, B.; Sauve, A. Sirtuin deacetylases as therapeutic targets in the nervous system. *Neurotherapeutics* **2013**, *10*, 605–620. [CrossRef] [PubMed]
15. Teng, Y.; Chen, Y.; Tang, X.; Wang, S.; Yin, K. PAD2: A potential target for tumor therapy. *Biochim. Et Biophys. Acta* **2023**, *1878*, 188931. [CrossRef]
16. Kulkarni, S.S.; Cantó, C. The molecular targets of resveratrol. *Biochim. Et Biophys. Acta* **2015**, *1852*, 1114–1123. [CrossRef]
17. Borra, M.T.; Smith, B.C.; Denu, J.M. Mechanism of human SIRT1 activation by resveratrol. *J. Biol. Chem.* **2005**, *280*, 17187–17195. [CrossRef] [PubMed]

18. Penantian, R.M.; Antarianto, R.D.; Hardiany, N.S. Effect of calorie restriction on the expression of sirtuin1 as an antiaging biomarker. *Makara J. Sci.* **2023**, *27*, 179–185.
19. Perinelli, D.R.; Mustafa, A.M.; Angeloni, S.; Sagratini, G.; Caprioli, G.; Cespi, M.; Bonacucina, G. Surface and emulsifying properties of raw ethanolic extracts from legumes containing soyasaponins and proteins. *Sustain. Chem. Pharm.* **2022**, *30*, 100872. [CrossRef]
20. Tsuno, Y.; Fujimatsu, T.; Endo, K.; Sugiyama, A.; Yazaki, K. Soyasaponins: A new class of root exudates in soybean (Glycine max). *Plant Cell Physiol.* **2018**, *59*, 366–375. [CrossRef] [PubMed]
21. Chitisankul, W.T.; Itabashi, M.; Son, H.; Takahashi, Y.; Ito, A.; Varanyanond, W.; Tsukamoto, C. Soyasaponin composition complexities in soyfoods relating nutraceutical properties and undesirable taste characteristics. *LWT* **2021**, *146*, 111337. [CrossRef]
22. Lin, X.; Zhang, C.; Huang, B. Hepatoprotective action mechanism and quantification of soyasaponin Bb in Abri Herba by HPLC and network pharmacology. *J. Ethnopharmacol.* **2025**, *337*, 118850. [CrossRef]
23. Chaturvedi, S.; Hemamalini, R.; Khare, S. Effect of processing conditions on saponin content and antioxidant activity of Indian varieties of soybean (*Glycine max* Linn.). *Ann. Phytomed.* **2012**, *1*, 62–68.
24. Jyothi, T.C.; Sindhu Kanya, T.C.; Appu Rao, A.G. Influence of germination on saponins in soybean and recovery of soy sapogenol I. *J. Food Biochem.* **2007**, *31*, 1–13. [CrossRef]
25. Tsai, C.-Y.; Chen, Y.-H.; Chien, Y.-W.; Huang, W.-H.; Lin, S.-H. Effect of soy saponin on the growth of human colon cancer cells. *World J. Gastroenterol.* **2010**, *16*, 3371–3376. [CrossRef] [PubMed]
26. Xie, Q.; Gu, X.; Chen, J.; Liu, M.; Xiong, F.; Wu, X.; Zhang, Y.; Chen, F.; Chen, H.; Li, M.; et al. Soyasaponins reduce inflammation and improve serum lipid profiles and glucose homeostasis in high fat diet-induced obese mice. *Mol. Nutr. Food Res.* **2018**, *62*, e1800205. [CrossRef] [PubMed]
27. Tang, Y.-Z.; Liu, Z.-Q. Free-radical-scavenging effect of carbazole derivatives on DPPH and ABTS radicals. *J. Am. Oil Chem. Soc.* **2007**, *84*, 1095–1100. [CrossRef]
28. Yu, J.; Ahmedna, M.; Goktepe, I. Effects of processing methods and extraction solvents on concentration and antioxidant activity of peanut skin phenolics. *Food Chem.* **2005**, *90*, 199–206. [CrossRef]
29. Azwanida, N.N. A review on the extraction methods use in medicinal plants, principle, strength and limitation. *Med. Aromat. Plants* **2015**, *4*, 196.
30. Floegel, A.; Kim, D.-O.; Chung, S.-J.; Koo, S.I.; Chun, O.K. Comparison of ABTS/DPPH assays to measure antioxidant capacity in popular antioxidant-rich US foods. *J. Food Compos. Anal.* **2011**, *24*, 1043–1048. [CrossRef]
31. Zhou, Z.; Fan, Z.; Meenu, M.; Xu, B. Impact of germination time on resveratrol, phenolic acids, and antioxidant capacities of different varieties of peanut (*Arachis hypogaea* Linn.) from China. *Antioxidants* **2021**, *10*, 1714. [CrossRef]
32. Choi, J.; Kim, J.; Lee, H.-D.; Cho, H.; Paje, L.A.; Shin, H.; Lee, S. Development of an analytical approach for the utilization of edible tree sprouts. *Nat. Prod. Sci.* **2022**, *28*, 27–32. [CrossRef]
33. Thaipong, K.; Boonprakob, U.; Crosby, K.; Cisneros-Zevallos, L.; Byrne, D.H. Comparison of ABTS, DPPH, FRAP, and ORAC assays for estimating antioxidant activity from guava fruit extracts. *J. Food Compos. Anal.* **2006**, *19*, 669–675. [CrossRef]
34. Tran, G.-H.; Uy, N.-P.; Kang, S.-H.; Heo, W.; Lee, E.-S.; Roh, S.-S.; Lee, S. HPLC/DAD analysis and antioxidant activity of adlay sprouts and seeds. *Separations* **2024**, *11*, 32. [CrossRef]
35. Lee, S.S.; Lee, S.M.; Kim, M.; Chun, J.; Cheong, Y.K.; Lee, J. Analysis of *trans*-resveratrol in peanuts and peanut butters consumed in Korea. *Food Res. Int.* **2004**, *37*, 247–251. [CrossRef]
36. Pae, S.-B.; Ha, T.-J.; Lee, M.-H.; Hwang, C.-D.; Shim, K.-B.; Park, C.-H.; Park, K.-Y.; Baek, I.-Y. Evaluation of characteristics of peanut sprout using Korean cultivars. *Korean J. Crop Sci.* **2011**, *56*, 394–399. [CrossRef]
37. Silva, C.G.; Monteiro, J.; Marques, R.R.N.; Silva, A.M.T.; Martínez, C.; Canle, L.M.; Faria, J.L. Photochemical and photocatalytic degradation of *trans*-resveratrol. *Photochem. Photobiol. Sci.* **2013**, *12*, 638–644. [CrossRef] [PubMed]
38. Zhao, Y.; Shi, M.; Ye, J.H.; Zheng, X.Q.; Lu, J.L.; Liang, Y.R. Photo-induced chemical reaction of trans-resveratrol. *Food Chem.* **2015**, *171*, 137–143. [CrossRef]
39. Santos Wagner, A.L.; Araniti, F.; Ishii-Iwamoto, E.L.; Abenavoli, M.R. Resveratrol exerts beneficial effects on the growth and metabolism of *Lactuca sativa* L. *Plant Physiol. Biochem.* **2022**, *171*, 26–37. [CrossRef] [PubMed]
40. Kim, S.H.; Yuk, H.J.; Ryu, H.W.; Oh, S.R.; Song, D.Y.; Lee, K.S.; Park, K.I.; Choi, S.W.; Seo, W.D. Biofunctional soyasaponin Bb in peanut (*Arachis hypogaea* L.) sprouts enhances bone morphogenetic protein-2-dependent osteogenic differentiation via activation of runt-related transcription factor 2 in C2C12 cells. *Phytother. Res.* **2019**, *33*, 1490–1500. [CrossRef] [PubMed]
41. Park, S.K.; Lee, H.L.; Kang, J.Y.; Kim, J.M.; Heo, H.J. Peanut (*Arachis hypogaea*) sprout prevents high-fat diet-induced cognitive impairment by improving mitochondrial function. *Sci. Rep.* **2022**, *12*, 6213. [CrossRef]
42. Kim, M.Y.; Jang, G.Y.; Lee, Y.J.; Kim, K.M.; Kim, D.J.; Lee, J.; Jeong, H.S. Application of high hydrostatic pressure for production of bioactive soyasaponin from black soybean [*Glycine max* (L.)] sprout. *J. Food Nutr. Res.* **2019**, *7*, 270–278.

43. Shimoyamada, M.; Okubo, K. Variation in saponin contents in germinating soybean seeds and effect of light irradiation. *Agric. Biol. Chem.* **1991**, *55*, 577–579.
44. Yoshiki, Y.; Kudou, S.; Okubo, K. Relationship between chemical structure and biological activities of triterpenoid saponins from soybean. *Biosci. Biotechnol. Biochem.* **1998**, *62*, 2291–2299. [CrossRef] [PubMed]

BioTech

MDPI

Article

Production of Multifunctional Hydrolysates from the *Lupinus mutabilis* Protein Using a *Micrococcus* sp. PC7 Protease

Keyla Sofía Llontop-Bernabé [1], Arturo Intiquilla [1,*], Carlos Ramirez-Veliz [1], Marco Santos [1], Karim Jiménez-Aliaga [1], Amparo Iris Zavaleta [1], Samuel Paterson [2] and Blanca Hernández-Ledesma [2,*]

1 Laboratory of Molecular Biology, Faculty of Pharmacy and Biochemistry, Universidad Nacional Mayor de San Marcos, Lima 01, Peru; keyla.llontop@unmsm.edu.pe (K.S.L.-B.); carlos.ramirez29@unmsm.edu.pe (C.R.-V.); kjimeneza@unmsm.edu.pe (K.J.-A.); azavaletap@unmsm.edu.pe (A.I.Z.)

2 Department of Bioactivity and Food Analysis, Institute of Food Science Research (CIAL, CSIC-UAM, CEI-UAM + CSIC), Nicolás Cabrera 9, 28049 Madrid, Spain; samuel.paterson@csic.es

* Correspondence: aintiquillaq@unmsm.edu.pe (A.I.); b.hernandez@csic.es (B.H.-L.)

Abstract: The growing demand for functional foods has driven the search for bioactive compounds derived from plant proteins. *Lupinus mutabilis* "Tarwi", a legume native to the Peruvian Andes, stands out for its high protein content and potential as a source of bioactive peptides (BPs). In this study, the functionality of the proteins contained in the albumin fraction (AF) isolated by tangential ultrafiltration (TFF) was investigated by using the OmicsBox software. The identified proteins were functionally classified into three groups: cellular component (35.57%), molecular function (33.45%), and biological process (30.97%). The isolated AF was hydrolysed with the native protease PC7 (HAP), optimizing the E/S ratio and time parameters. Additionally, sequential hydrolysis of the PC7 protease and alcalase (HAPA) was performed. In vitro multifunctionality assays, HAP and HAPA demonstrated the ability to scavenge radicals (ABTS and ORAC) and inhibit angiotensin-converting enzyme (ACE)-I and dipeptidyl peptidase IV (DPP-IV). The findings of this study highlight the potential of *L. mutabilis* albumin hydrolysate as a multifunctional ingredient for functional foods aimed at managing chronic conditions associated with oxidative stress, hypertension, and/or metabolic disorders.

Keywords: *Lupinus mutabilis*; albumin fraction; ultrafiltration; protease PC7; multifunctionality

Key Contribution: This is the first report where a neutral protease is used for the production of protein hydrolysates from *L. mutabilis* seeds, with antioxidant, ACE-I, and DPP-IV inhibitory properties.

1. Introduction

In recent years, changes in human diet behaviour have boosted the consumption of functional foods, with an estimated annual growth of 8.5% until 2030 [1]. Plant-based proteins offer an alternative source of bioactive compounds, presenting advantages over animal-based proteins due to their diversity, cost-effectiveness, large-scale production, and environmental sustainability [2]. Legumes stand out among plant sources for their nutritional value, as they constitute a source of high-quality proteins, amino acids (AAs), and minerals [3]. Beyond their nutritional contribution, legumes contain a variety of bioactive compounds, such as isoflavones, bioactive peptides (BPs), saponins, phenolic compounds, galactosides, and anti-nutritional factors [4,5]. While the latter, such as phytic acid and alkaloids, have traditionally been associated with reduced nutrient bioavailability, emerging evidence highlights their potential bioactive properties, including antioxidant and antidiabetic effects [6,7]. These bioactive components exhibit metabolic, hormonal, and digestive regulatory effects, in addition to prebiotic properties that vary depending on the specific compound type, as well as digestibility and bioavailability [8–10].

BPs are short sequences of AAs released from the native protein through in vitro enzymatic hydrolysis, fermentation, and/or gastrointestinal digestion. BPs primarily can exhibit antioxidant, antihypertensive, and antidiabetic properties [11]. Their biological activity is determined by the AA sequence, the type of terminal AA, the length of the chain, the electrical charge, and the hydrophobic and hydrophilic properties, among others [12]. In particular, the aromatic (tryptophan, tyrosine, and phenylalanine), imidazole ring (histidine), and sulphur (cysteine and methionine) AAs present in BPs have been recognized to contribute to their ability to donate hydrogen or electrons, inhibiting the formation of free radicals [9,13]. This mechanism is crucial for the prevention of metabolic and chronic diseases associated with oxidative stress, such as obesity, type II diabetes, neurodegenerative disorders, and cancer [12].

Studies have reported that legume proteins such as *Lupinus albus* and *L. angustifolius* exhibit antioxidant, antihypertensive, and antidiabetic activity primarily after being treated with commercial enzymes such as pepsin, pancreatin, flavourzyme, neutrase, and alcalase [14–17]. Although native proteases from non-commercial bacterial sources have been investigated for their ability to release antioxidant peptides [18], their application to legume proteins has not yet been explored. In this regard, the PC7 protease from *Micrococcus* sp., isolated from *Pilluana salterns* in Peru, represents a promising biocatalyst for generating unique BPs from Andean legumes, as demonstrated in the study by Bautista et al. [19]. This protease operates at pHs (7.0 to 8.0) and temperatures (25 to 40 °C) that avoid adverse handling conditions, favouring its scaling. This prevents excessive protein denaturation, preserving the functional properties of the released peptides. To enhance peptide yield and reduce hydrolysis time and process cost, it is essential to optimize the hydrolysis conditions based on the parameters of temperature, pH, reaction time, and enzyme concentration.

L. mutabilis "Tarwi", a legume native to the Peruvian Andes, is notable for its high protein content, ranging from 32.0 to 52.6% (*w*/*w*), primarily composed of albumin and globulin fractions. Traditionally, lupin storage proteins are also known as α-, β-, γ-, and δ-conglutins [20]. The growing interest in these proteins is not only due to nutritional reasons but also because they are a source of bioactive compounds. A total of 25 peptides with antioxidant, antihypertensive, and antidiabetic capacities have been identified in a protein hydrolysate of *L. mutabilis*; however, the exact precursor protein is unknown [21]. The albumin fraction is one of the proteins with the greatest potential for industrial development due to its high solubility in pure water, which favours its extraction and processing, making it a valuable ingredient for the formulation of healthy products [20,22]. The quality and functionality of these proteins depend to a large extent on the extraction method used. In this sense, tangential flow ultrafiltration (TFF) is one of the most efficient techniques, as it allows soluble proteins to be isolated in their native state while preserving their structure and functionality [23].

Proteomic analysis based on high-resolution mass spectrometry is a powerful tool for optimizing protein characterisation. This technique allows the identification, quantification, and analysis of proteins, even those found at low concentrations, allowing the study of their main functions in plants [24,25]. Previous proteomic studies on *Lupinus* spp. have focused on taxonomic identification, allergen detection, and conglutin diversity characterization [24–26].

This highlights the importance of expanding proteomic research on the albumin of *L. mutabilis* to explore its potential as a source of BPs for functional food formulations.

In this context, this study aimed to harness the albumin fraction (AF) proteins of *L. mutabilis*—characterized for the first time via proteomic analysis—to develop a multifunctional protein hydrolysate with in vitro antioxidant, antihypertensive, and antidiabetic properties. To achieve this, we employed the novel native PC7 protease (HAP), previously undescribed for this application. Furthermore, sequential hydrolysis using PC7 protease followed by alcalase (HAPA) significantly enhanced these bioactive properties. Beyond advancing scientific understanding, this innovative approach enhances the nutritional and functional value of *L. mutabilis* seeds while promoting their economic valorisation and the creation of novel food ingredients.

2. Materials and Methods

2.1. Materials and Reagents

The seeds of *L. mutabilis* "Tarwi" were collected in Huamachuco (Sanchez Carrion-La Libertad, Peru) in 2023.

Alcalase 2.4 L (*Bacillus licheniformis*, >2.4 U/g) was obtained from Sigma-Aldrich (Bagsvaerd, Denmark), 2,2′-azino-bis (3-ethylbenzothiazoline-6-sulfonic acid) (ABTS), 6-hydroxy-2,5,7,8-tetramethylcroman-2-carboxylic acid (Trolox), fluorescein disodium (FL), 2,2′-azobis dihydrochloride (2-amidinopropane) (AAPH), and the dipeptidyl peptidase IV (DPP-IV) assay kits were obtained from Cayman Chemicals (Ann Arbor, MI, USA), and orthophthalaldehyde (OPA) was obtained from Merck Millipore Corp (Darmstadt, Germany). All other chemicals were of analytical grade.

2.2. Production of PC7 Protease

The strain *Micrococcus* sp. PC7 was provided by the Molecular Biology Laboratory, Faculty of Pharmacy and Biochemistry, Universidad Nacional Mayor de San Marcos (Lima, Peru). The culture was grown in a medium containing 10 g/L glucose, 5 g/L yeast extract, and 5% salt water, adjusted at pH 8.0 [27]. The strain was inoculated into the medium at an optical density $(OD)_{600}$ of 0.5, using 10% (v/v) of the culture in 400 mL of medium. Incubation was carried out at 30 °C, 150 rpm, for 48 h. After incubation, cells were removed by centrifugation at 8000× g for 20 min, and the resulting supernatant was collected as the source of proteases, referred to as the crude extract. This crude extract was concentrated to a final volume of 10 mL using a 10 kDa molecular weight cut-off (MWCO) membrane (Millipore, CA, USA), at 7500× g for 20 min at 4 °C (Bautista et al., 2024 [19]).

The proteolytic activity of the crude extract was determined following the method described by Bautista et al. [19]. One unit of enzyme activity (U) was defined as the amount of enzyme required to produce a 0.1 increase in absorbance at 280 nm, under the specified test conditions. Specific activity was expressed as U/mg of protein.

2.3. Obtention of Albumin Fraction from L. mutabilis Seeds

The albumin fraction (AF) was extracted from *L. mutabilis* seeds following the method described by Sironi et al., with minor modifications [28]. Defatted flour was dispersed in Milli-Q water 1:9 (w/v), homogenized, and adjusted to pH 7.0 using 1 M NaOH. The suspension was stirred with a magnetic stirrer (VWR 4X4, PA, USA) at 1000 rpm for 1 h at 24 °C. Subsequently, the suspension was centrifuged at 2700× g for 25 min at 10 °C and the supernatant was collected.

The protein extract underwent TFF using a Pellicon® 3 cassette system with an Ultracel® regenerated cellulose membrane (nominal MWCO of 3 kDa, total area 0.11 m^2; Millipore corporation, Billerica, MA, USA). The process was performed at an initial transmembrane pressure (TMP) of 24 psi with a feed flow rate of 450 mL/min. The TFF continued until the volume was reduced to 200 mL, achieving a volumetric concentration factor (VCF) of 5.75. Both retentate and permeate samples were collected for subsequent analyses. The retentate, representing the concentrated AF, was freeze-dried, and stored at 4 °C until further use. The protein content was determined using the bicinchoninic acid (BCA) method, following the instructions of the Pierce™ BCA Protein Assay Kit (Thermo Fisher Scientific, Waltham, MA, USA). Bovine serum albumin (BSA, 25–1000 µg/mL) was used as the standard.

2.4. Characterization of the Albumin Fraction

2.4.1. Determination of Protein and Amino Acid Content

The protein concentration of AF was also measured via the Kjeldahl method [29], employing a block digester (J.P. Selecta, Barcelona, Spain) and a Buchi Kjeldahl K-314 distillation unit (BÜCHI Labortechnik AG, Flawil, Switzerland). The analysis was carried out in duplicate using a conversion factor of 5.71. Additionally, the amino acid content was measured in duplicate using a Biochrom 30 Series Amino Acid Analyser (Biochrom, Cambridge, UK), following the previously internationally established methodology [30].

2.4.2. In-Gel Digestion (Stacking Gel)

The AF was suspended in 50 µL of sample buffer and loaded into 1.2 cm-wide wells of a standard SDS-PAGE gel (0.75 mm thick, 4% stacking, 10% resolving; Bio-Rad, Hercules, CA, USA) following a previously described protocol [31]. The run was halted once the front migrated 3 mm into the resolving gel, concentrating the entire proteome at the stacking/resolving gel interface. Unseparated protein bands were stained with Coomassie dye (Bio-Rad), excised, cut into 2 × 2 mm cubes, and transferred to 0.5 mL microcentrifuge tubes [32]. The gel fragments were destained using an ACN/H_2O (1:1, *v*/*v*) solution, then subjected to reduction and alkylation—cysteinyl disulfide bonds were reduced with 10 mM 1,4-dithiothreitol (DTT) at 56 °C for 1 h, followed by alkylation with 10 mM iodoacetamide at room temperature for 30 min in the dark. In situ digestion was carried out using sequencing-grade trypsin (Promega, Madison, WI, USA) as per Shevchenko et al. [33]. The gel pieces were dehydrated with acetonitrile (ACN), which was subsequently removed, and then dried in a speedvac before being rehydrated in 100 mM Tris-HCl (pH 8.0), 10 mM $CaCl_2$ containing 60 ng/mL trypsin at a 5:1 protein/enzyme (*w*/*w*) ratio. The samples were kept on ice for 2 h and incubated at 37 °C for 12 h. Digestion was terminated with the addition of 1% trifluoroacetic acid (TFA). The resulting supernatants were dried and desalted using OMIX Pipette Tips C18 (Agilent Technologies, Santa Clara, CA, USA) before analysis. The digestion process was carried out in the presence of 0.2% RapiGest (Waters, Milford, MA, USA).

2.4.3. Reverse-Phase Liquid Chromatography (RP-LC-MS/MS) Analysis (Dynamic Exclusion Mode)

The desalted protein digest was dried, reconstituted in 10 µL of 0.1% formic acid, and analysed using RP-LC-MS/MS on an Easy-nLC 1200 system coupled to an LTQ-Orbitrap Velos Pro hybrid mass spectrometer (Thermo Scientific, Waltham, MA, USA). Peptides were first concentrated online via reverse-phase chromatography with a 0.1 mm × 20 mm C18 RP precolumn (Thermo Scientific) and then separated on a 0.075 mm × 250 mm bioZen 2.6 µm Peptide XB-C18 RP column (Phenomenex, Torrance, CA, USA) at a flow rate of 0.25 µL/min. Elution was carried out over 180 min using a dual-gradient system: 5–25% solvent B over 135 min, 25–40% over 45 min, 40–100% over 2 min, followed by 100% solvent B for 18 min (solvent A: 0.1% formic acid in water; solvent B: 0.1% formic acid, 80% ACN in water). Electrospray ionization (ESI) was performed using a Nano-bore stainless steel emitter (30 µm ID; Proxeon, Odense, Denmark) with a spray voltage of 2.1 kV and an S-Lens setting of 60%. The Orbitrap resolution was set at 30,000 [34]. Peptides were detected in survey scans ranging from 400 to 1600 amu (1 µscan), followed by twenty data-dependent MS/MS scans (Top 20) with a 2 u isolation width (*m*/*z* units), a normalized collision energy of 35%, and dynamic exclusion for 60 s. Charge-state screening was enabled to discard unassigned and singly charged protonated ions. The mass spectrometry proteomics data have been deposited to the ProteomeXchange Consortium via the PRIDE [35] partner repository with the dataset identifier PXD062562 and 10.6019/PXD062562.

2.4.4. Data Processing

Peptide identification from raw data was conducted using the PEAKS Studio v11.5 search engine (Bioinformatics Solutions Inc., Waterloo, ON, Canada). The database search was performed against uniprot-phaseoleae.fasta (480,563 entries; UniProt release 02/2024) or a combined database consisting of uniprot-lupinus-albus, uniprot-lupinus-angustifolius, and uniprot-lupinus-mutabilis.fasta (69,007 entries; UniProt release 06/2024) using a decoy-fusion database. Search parameters included semi-specific tryptic cleavage at arginine and lysine, allowance for up to two missed cleavage sites, and mass tolerances of 20 ppm for precursor ions and 0.6 Da for MS/MS fragment ions. Modifications considered in the search included optional methionine oxidation and cysteine carbamidomethylation. The false discovery rate (FDR) was restricted to 0.01 for both peptide–spectrum matches (PSMs) and protein identifications. Proteins were considered confidently identified only if at least two unique peptides were detected in LC-MS/MS analyses [34,36].

2.4.5. Proteomic Functional Analysis

A proteomic functional analysis of all the detected proteins from AF was carried out through the BLAST2GO methodology [37] using the OmicsBox 3.4 bioinformatic software (Biobam, Valencia, Spain). Three distinct functional groups were made: cellular component, molecular function, and biological process. To complete the alignment, the non-redundant database for protein sequences was used using the taxonomy filter: *Phaseoleae* (code 163735), *L. albus* (code 3870), *L. angustifolius* (code 3871), and *L. mutabilis* (code 53232). The step by step full workflow followed for the functional analysis is shown in Supplementary Figure S1.

2.5. *Hydrolysis of Albumin Fraction with PC7 Protease*

An aqueous suspension of the AF (65 mg/mL) was prepared and heated to 85 °C for 10 min before hydrolysis with PC7 protease. The reaction conditions were maintained at pH 8.0, a temperature of 40 °C, and 150 rpm agitation [19]. The E/S ratio and hydrolysis time (T) were defined using a central composite design (CCD). The E/S ratio ranged from 44 to 256 U/mg of substrate (X_1), while the hydrolysis time ranged from 26.4 to 153.6 min (X_2). The dependent variables were the degree of hydrolysis (*DH*) and the antioxidant capacity (ABTS). A total of 11 assays were performed, based on five levels of the independent variables, including 4 factorials, 4 axial points, and 3 replicates at the central point. The experimental design was generated using the Statistica v10.0 software.

The regression models describing the relationship between the dependent variables (Y) and independent variables (X) were expressed using the following polynomial equation:

$$Y = \beta_0 + \beta_1 X_1 + \beta_2 X_2 + \beta_3 X_1^2 + \beta_4 X_2^2 + \beta_5 X_1 X_2 \quad (1)$$

Enzymatic hydrolysis was terminated by heating the reaction at 100 °C for 10 min, followed by cooling and centrifugation at 5000× *g* for 10 min at 4 °C. The resulting supernatants were freeze-dried and stored at −20 °C for subsequent use.

The DH was determined based on the method described by Nielsen et al. by measuring the free amino groups (h) by reaction with OPA and expressed as mg L-serine/mL [38]. DH (%) was calculated using the formula:

$$\%DH = \frac{h}{ht} \times 100 \quad (2)$$

The total amino group content (ht = 10.15) was determined in samples subject to complete hydrolysis (100%) with 6 N HCl at 121 °C for 1 h. Soluble protein concentrations in the hydrolysates were quantified using the BCA assay as outlined in Section 2.3.

2.6. Sequential Hydrolysis of Albumin Fraction with Protease PC7 and Alcalase (HAPA)

The AF was sequentially hydrolysed using PC7 protease at a ratio of 168 U/mg substrate, under conditions of pH 8.0 and 40 °C, for 105 min at 150 rpm (HAP). This was followed by the hydrolysis with alcalase at an E/S ratio of 1:50 (w/w), pH 8.5, and 50 °C for 2 h [39]. The reaction was terminated by heating at 100 °C for 10 min, followed by centrifugation at 5000× g for 10 min at 4 °C. The resulting supernatant (HAPA) was freeze-dried and stored at −20 °C for subsequent use.

2.7. SDS-PAGE Electrophoretic Profile

The electrophoretic profiles of AF, HAP, and HAPA were analysed using SDS-PAGE-Tricine, under reducing and non-reducing conditions, following the method described by Haider et al., with slight modifications [40]. Samples were treated with sample buffer containing 1% (w/v) sodium dodecyl sulphate (SDS), 24% glycerol, and 0.02% (w/v) Coomassie brilliant Blue in 0.1 mM Tris-HCl buffer at pH 6.8. For reducing conditions, 4% (v/v) 2-mercaptoethanol was added. The samples were incubated at 65 °C for 10 min. Protein aliquots (10 µg) were loaded onto a resolving gel (10% acrylamide) with a stacking gel (4% acrylamide). Electrophoresis was conducted in a vertical gel unit (Mini PROTEAN, Bio-Rad) at 80 V for 20 min, followed by 100 V for 2 h. An MW marker ranging from 10 to 225 kDa (Perfect Protein Marker, Sigma-Aldrich) was used.

2.8. Determination of Biological Activities

2.8.1. ABTS Assay

The $ABTS^{•+}$ radical scavenging activity was measured according to the method of Re et al., with minor modifications [41]. The 7 mM $ABTS^{•+}$ radical was generated by mixing 3 mL of 10 mM ABTS with 3 mL of 2.45 mM potassium persulfate. The reaction mixture was protected from light by storing it in an amber glass vial and was allowed to stand for 16 h at 23 °C. Subsequently, the solution was diluted with 5 mM PBS (pH 7.4) to achieve an absorbance of 0.70 ± 0.02 at 734 nm. Trolox was used as a standard at concentrations ranging from 2 to 25 µM. For the assay, 180 µL of $ABTS^{•+}$ solution was mixed with 20 µL of PBS (blank), Trolox (standard), or sample (hydrolysate), and the absorbance was measured at 734 nm after 5 min incubation at room temperature, using a Tecan Infinite 200PRO plate reader (Tecan Group AG, Männendorf, Switzerland), controlled by the Icontrol software version 1.11.10. The results were expressed as µmol of Trolox equivalents (TEs) by g of protein. Samples were analysed in triplicate.

2.8.2. Oxygen Radical Absorbance Capacity (ORAC) Assay

The ORAC assay was performed as described by Hernández-Ledesma et al., with slight modifications [42]. Briefly, a 200 µL reaction mixture containing disodium fluorescein (116.61 nM), 14 mM AAPH, and either Trolox (0–3 nM) or sample (hydrolysate) was prepared in 75 mM PBS (pH 7.4). The mixture was incubated at 37 °C, and fluorescence was measured at the excitation wavelength of 485 nm and emission wavelength of 520 nm every 2 min over 120 min using a plate reader Tecan Infinite 200PRO (Tecan Group AG). The values were expressed as μmol TE/mg of protein. Samples were analysed in triplicate.

2.8.3. Angiotensin-Converting Enzyme (ACE) Inhibitory Activity

The angiotensin-converting enzyme (ACE) inhibitory activity was determined using the methodology described by Hayakari et al., with slight modifications [43]. Briefly, a 200 μL mixture containing 1 mU ACE, 1.25 mM hippuryl histidyl-leucine, 0.05 μM Captopril (control), or sample (hydrolysate) was prepared and incubated at 37 °C for 60 min. The reaction was terminated by heating the samples at 100 °C for 1 min. Subsequently, 100 μL of a 3% (*v*/*v*) trichloro-5-triazine/dioxane solution was added, and the mixture was stirred and centrifuged at 1000× *g* for 10 min. Finally, absorbance was measured at 382 nm using the Tecan Infinite 200PRO (Tecan Group AG) plate reader. The results were expressed as sample concentration required to inhibit ACE activity by 50% (IC_{50}). Samples were analysed in triplicate.

2.8.4. Dipeptidyl Peptidase-IV (DPP-IV) Inhibitory Activity

The antidiabetic activity of the hydrolysates was assessed by evaluating their DPP-IV inhibitory activity [44]. A 100 μL reaction mixture was prepared, consisting of 5 μL of enzyme solution, 5 μL of substrate, sitagliptin (standard), or samples at various concentrations. The mixture was incubated at 37 °C for 30 min. Fluorescence measurements were recorded every 2 min (λ excitation = 360 nm/λ emission = 460 nm) using the Tecan Infinite 200PRO (Tecan Group AG) plate reader. The results were expressed as the sample concentration required to inhibit 50% of DPP-IV activity (IC_{50}). Samples were analysed in triplicate.

2.9. Statistical Analysis

All experiments were performed in triplicate and presented as mean ± standard deviation. The data were subjected to a one-way analysis of variance (one-way ANOVA) followed by Tukey's test. Differences were considered statistically significant when $p < 0.05$. Figures were prepared using GraphPad Prism 8 software. Analysis and interpretation of data on hydrolysis optimization were performed using Statistica v10.0 software.

3. Results and Discussion

3.1. Obtention of Albumin Fraction (AF) from Lupinus mutabilis by Tangential Flow Ultrafiltration

The AF is a water-soluble, sulphur-rich protein (δ-conglutin) that performs storage and regulatory functions in seeds [45]. TFF was employed as an alternative to isoelectric precipitation for AF isolation. This technique facilitates the fractionation of the target protein while preserving its native structure and functionality [46]. The work design is shown in Figure 1A and the operating parameters are given in Supplementary Table S1.

The volumetric flow rate remained relatively stable throughout the 40 min process. This stability was counterbalanced by a slight increase in the TMP, which rose from 24.0 ± 0.35 to 26.13 ± 1.94 psi, without a statistically significant reduction ($p < 0.05$) in the normalized water permeability (NWP) parameter. However, a total protein loss of 16.48 ± 0.49% was detected (Figure 1B). This loss could be attributed to permeate entrainment phenomena and protein–membrane interactions, both of which contribute to protein loss and are linked to the fouling index. Membrane fouling is shown in Figure 1C. No significant accumulation of fouling material was detected on the membrane until 30 min after the process began. Subsequently, a slight increase in fouling was observed, reaching 3.03 ± 1.43% after 40 min. This trend might be explained by the increase in protein concentration, which promotes concentration polarization on the membrane surface and partial pore blockage as the process progresses [47]. These findings underscore the importance of optimizing the UF process by carefully selecting membranes and controlling operating parameters to achieve desired yields while minimizing fouling and polarization effects.

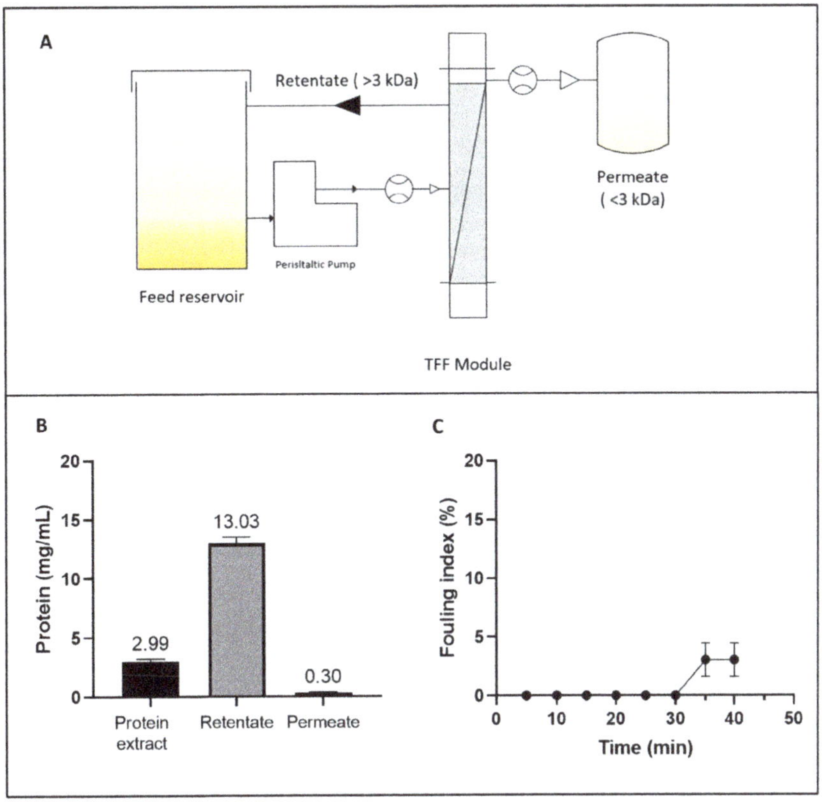

Figure 1. (**A**) Diagram of the tangential flow ultrafiltration (TFF) system for concentrating the albumin fraction (AF) from *Lupinus mutabilis*. (**B**) Protein concentration in the initial extract, retentate, and permeate during the TFF process. (**C**) Fouling index (%) of TFF as a function of time.

The TFF approach enabled a 4.36-fold concentration of albumin protein (Figure 1B), achieving a VCF of 5.75, and yielding a 2.9-fold (75.82% *w*/*w*) increase in recovery compared to the isoelectric precipitation process (26.23%) reported in a previous study [15]. The improvement of the yield (%) could be attributed to the ability of TFF to separate and concentrate proteins in a controlled environment, minimizing protein loss during the process. Similarly, Hojilla et al. demonstrated the efficacy of UF combined with diafiltration (DF), achieving a 72% recovery rate in *L. albus* using a 5 kDa polyether sulfone membrane, with a flow rate of 2.7 L/h [48]. These findings align with the results obtained in the current study, reinforcing the notion that membrane-based technologies are versatile and effective tools for processing plant-derived proteins.

The protein content of the resultant AF was 82.87 ± 0.21%. Its AA composition is shown in Table 1. The chromatogram obtained from the analysis is shown in Supplementary Figure S2.

Table 1. Amino acid composition (g/100 g protein and g/100 g biomass) of albumin fraction (AF) from *Lupinus mutabilis*.

Amino Acid	AF from *L. mutabilis*	
	g/100 g Protein	g/100 g Biomass
Essential (EAA)		
Lysine (Lys)	3.52 ± 0.05	2.92 ± 0.04
Phenylalanine (Phe)	2.38 ± 0.03	1.97 ± 0.03
Tyrosine (Tyr)	2.06 ± 0.11	1.71 ± 0.09
Methionine (Met)	0.34 ± 0.03	0.28 ± 0.03
Cysteine (Cys)	1.42 ± 0.12	1.17 ± 0.10
Threonine (Thr)	2.38 ± 0.04	1.97 ± 0.03
Leucine (Leu)	3.73 ± 0.02	3.09 ± 0.02
Isoleucine (Ile)	1.91 ± 0.07	1.58 ± 0.06
Valine (Val)	1.70 ± 0.04	1.41 ± 0.03
Non-essential (NEAA)		
Aspartic acid + Asparagine (Asx)	6.41 ± 0.03	5.31 ± 0.03
Glutamic acid + Glutamine (Glx)	20.01 ± 0.26	16.58 ± 0.21
Serine (Ser)	3.63 ± 0.06	3.01 ± 0.05
Histidine (Hys)	2.07 ± 0.05	1.72 ± 0.04
Arginine (Arg)	8.54 ± 0.29	7.08 ± 0.24
Alanine (Ala)	2.31 ± 0.02	1.91 ± 0.01
Proline (Pro)	2.91 ± 0.00	2.41 ± 0.00
Glycine (Gly)	2.55 ± 0.02	2.11 ± 0.02
EAA	19.44	16.10
NEAA	48.43	40.13
TAA	67.87	56.23
EAA × 100/TAA (%)	28.64	
EAA × 100/NEAA (%)	40.14	
HAA × 100/TAA (%)	27.64	
AAA × 100/TAA (%)	6.54	

EAA: essential amino acids; NEAA: non-essential amino acids; TAA: total amino acids; HAA: hydrophobic amino acids (Ala + Val + Ile + Leu + Tyr + Phe + Trp + Met + Pro + Cys); AAA: aromatic amino acids (Phe + Trp + Tyr).

The amino acid Trp was not detected according to the official AOAC International method [30]. This process was performed with 6 N HCl at high temperatures (110 °C) for 24 h to break peptide bonds. The indole ring of Trp is highly susceptible to degradation under acidic conditions, forming undetectable non-volatile compounds in the final hydrolysate. Among the essential AAs (EAAs), Leu and Lys were the most abundant, with values of 3.73 ± 0.02 and 3.52 ± 0.05 g/100 g of protein, respectively. Asp + Asn, Arg, and Glu + Gln were the most abundant non-essential AAs (NEAAs), with values of 20.01 ± 0.26, 8.54 ± 0.29, and 6.41 ± 0.03 g/100 g of protein, respectively. The ratio of EAAs to total AAs (TAAs) was 29%, and the ratio of EAAs to NEAAs was 40%. Both values were lower than the protein reference pattern (EAAs/TAAs, 40%; EAAs/NEAAs, 60%) raised by FAO/WHO.

Figure 2 illustrates the electrophoretic profile under reducing (β-mercaptoethanol) and non-reducing conditions of FA obtained by TFF. Under reducing conditions, bands with estimated molecular weights of 10, 14, 20, and 24 kDa were observed. Under non-reducing conditions, protein bands of 16 and 18 kDa were additionally detected. These findings were consistent with those reported by Nadal et al. for *L. luteus*, where bands around 20 kDa and below 15 kDa were observed [49]. Similarly, Muranyi et al. identified an albumin with an MW of approximately 13 kDa in *L. angustifolius* seeds [50]. Salmanowicz further noted that the AF consisted of polypeptide bands containing four to eight isoforms, characterized by two subunits of 3–6 kDa and 8–12 kDa, linked by four disulfide bridges, with variations depending on the species [51,52]. These observations were supported by Foley et al., who highlighted genetic variability within the same species, influenced by the seed development stage [53].

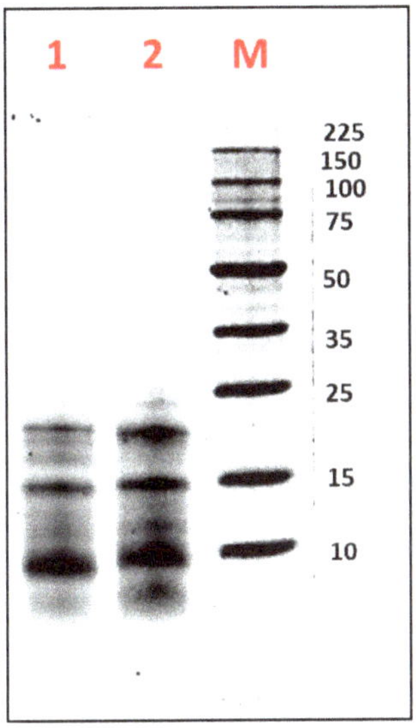

Figure 2. Electrophoretic analysis (SDS-PAGE-Tricine) of the albumin fraction (AF) from *Lupinus mutabilis* "Tarwi" seeds obtained by tangential flow ultrafiltration (TFF), under (1) non-reducing conditions and (2) reducing conditions. M: molecular weight marker (Perfect Protein Marker 10-225 kDa). A total of 10 µg of protein was loaded per sample.

3.2. Proteomic Characterization and Functional Analysis of L. mutabilis Albumin

Following RP-LC-MS/MS analysis of the AF sample and data processing using the PEAKS and SPIDER search tools [54], a total of 2424 peptide–spectrum matches (PSMs) and 2474 scans (representing MS/MS scans associated with peptide spectra) were identified. Additionally, 955 features were detected through database searches alone, 887 peptide sequences contained modifications excluding isoleucine/leucine differentiation, and 564 features were identified without modifications or isoleucine/leucine differentiation. Furthermore, the PEAKS software detected a total of 242 proteins following the database search and RP-LC-MS/MS analysis. Since homology based protein identification relies on sequence data, PEAKS organizes proteins into groups that include all proteins identified with the same peptides [31,55]. In our case, 111 protein groups were identified, and, out of the 242 proteins detected, 168 proteins were reliably identified with two or more unique peptides, and 74 proteins had less than two unique peptides.

The proteomic features and details of proteins with at least two unique peptides detected in the AF sample, were collected and are made available in Supplementary Table S2. The AF constitutes approximately 10% of the total protein content in *L. mutabilis* seeds. Within this fraction, while some proteins, such as δ-conglutin, are well-characterized components, others have not yet been comprehensively described in the available literature [56]. Therefore, further proteomic studies would be necessary to identify and quantify the specific proteins present in this fraction. Concerning the proteomic functional analysis, all 242 proteins detected in the sample were used. This strategy made it possible to detect proteins that might be missed or underrepresented due to technical challenges, such as low concentration, inefficient extraction, or poor ionization during MS analysis [57]. By incorporating these proteins, even those with barely detectable or missing peptides were accounted for, offering new biological and functional insights—particularly valuable for less-studied species like *L. mutabilis* [58]. Furthermore, the 74 proteins identified with fewer than two unique peptides may still play key roles in cellular functions or metabolic pathways but are often excluded in standard proteomics studies. Examining the full dataset allows researchers to highlight proteins with potential bioactive properties, nutritional significance, or industrial relevance that might otherwise go unnoticed. The classification of these proteins at functional levels 1 and 2 is depicted in Figure 3, with a comprehensive breakdown available in Supplementary Figures S2–S4.

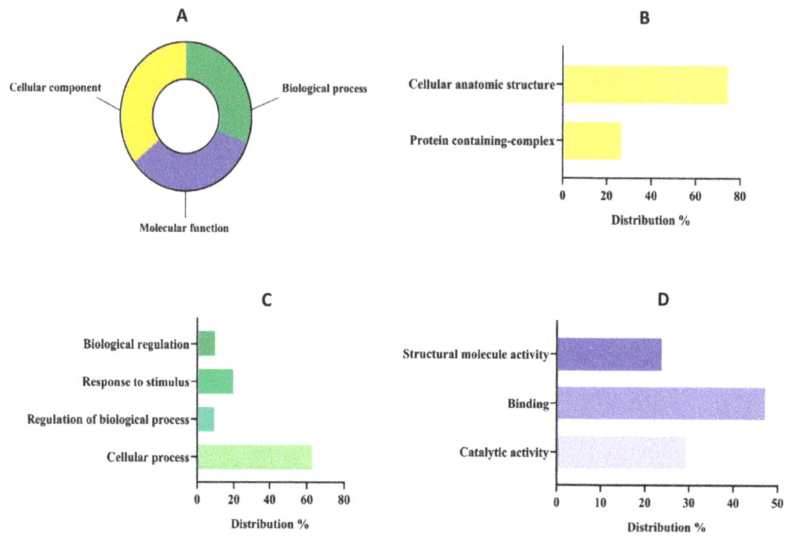

Figure 3. Functional classification of detected proteins from the albumin fraction (AF) of *Lupinus mutabilis* "Tarwi" based on gene ontology (GO). (**A**) Proteomic percentage distribution at GO level 1 classification. (**B**) Molecular function GO level 2 distribution. (**C**) Biological process GO level 2 distribution. (**D**) Cellular component GO level 2 distribution.

The distribution of proteins at level 1 was quite similar (Figure 3A) among groups since 201 sequences were classified as cellular components (35.57%), 189 sequences into molecular function (33.45%), and 175 sequences into the biological process group (30.97%). The group of cellular components might reunite the major number of sequences because the primary function of the AF (mainly composed of 2S albumins) is to accumulate in specialized cellular compartments serving as a structural and AA reservoir [59]. In fact, concerning the cellular component group (Figure 3B), most of the proteins were classified within the cellular anatomic structure group (73.89%) followed by the protein-containing complex group (26.10%). These results were in tune with the main functions of albumin-related proteins, as these proteins are primarily localized in protein storage vacuoles (PSVs) which are specialized organelles in seed cells dedicated to storing reserve proteins. Within PSVs, 2S albumins contribute to the formation of protein aggregates or complexes, ensuring the efficient storage and stability of these reserves during seed dormancy [60]. Upon germination, these complexes are mobilized to provide EAAs for the development of seeds [61].

Regarding the biological process group (Figure 3C), the analysis revealed that most of them were related to cellular processes (62.40%), followed by a response to stimulus (19.37%), biological regulations (9.3%), and regulation of biological processes (8.91%). Our results are aligned with previous works about albumin implications in biological processes, as it plays a crucial role in seed viability, growth, and early plant development and its expression may be modulated in response to environmental stresses such as drought or nutrient deficiency [26]. Moreover, albumins have been reported to contribute to the osmotic balance within seed cells by maintaining cellular turgor and stability, which is vital for preserving cell structure during the desiccation and rehydration phases [62]. Finally, within the sequences associated with molecular function at level 2 (Figure 3D), most of them were related to binding (47.03%), followed by catalytic activities (29.26%) and structural molecule activities (23.69%).

3.3. Optimization of Albumin Hydrolysis with Protease PC7

In a previous study, the PC7 protease, derived from *Micrococcus* sp. PC7, was isolated and characterized as effective in hydrolysing proteins from different legumes [19]. In the present study, the PC7 protease was concentrated using UF with a 10 kDa MWCO. The volumetric and specific activity of the concentrate protease was 4080 U/mL and 1843.7 U/mg of protein, respectively.

The CCD was applied to optimize both the *DH* and antioxidant activity of HAP by determining the optimal E/S ratio and hydrolysis time. The range of analysis for the E/S ratio was 75–225 U/mg of substrate, and the reaction time was 45–135 min (Table 2). Proteolysis was assessed by the *DH* that was determined by measuring the free amino groups produced during hydrolysis. Values ranging from 18.93% to 39.86% were obtained. The highest values (*DH* > 35%) were reached under conditions with the highest E/S ratios (225–256 U/mg) and longest reaction times (135–153.6 min), as evidenced by treatments 4, 6, and 8. These results could be due to the high enzyme concentration and long incubation times that facilitated the interaction between enzyme and substrate, promoting the bond cleavage. A similar behaviour was observed in the treatments corresponding to the central points (9–11) of E/S (150 U/mg) and times (90 min), achieving a *DH* between 29% and 33%. In contrast, the lowest *DH* values (18.93–26.39%) were associated with conditions involving low E/S ratios (44–75 U/mg) and short reaction times (26.4–45 min), as seen in treatments 1, 3, 5, and 7.

Table 2. Observed and predicted values for degree of hydrolysis (*DH*%) and ABTS$^{\bullet+}$ radical scavenging activity (TEAC) from the experimental design optimizing enzyme/substrate ratio (E/S) and hydrolysis time (T) of the albumin fraction (AF) from *Lupinus mutabilis* hydrolysed with protease PC7 (HAP).

Run	X1 E/S (U/mg)	X2 Time (min)	Y1 *DH* (%) Observed	Predicted	Y2 TEAC (µmol TE/g Protein) Observed	Predicted
1	75	45	22.57 ± 0.35	18.94	142.3 ± 7.26	136.9
2	225	45	32.91 ± 0.74	23.55	193.3 ± 0.19	185.4
3	75	135	25.14 ± 0.93	29.40	195.0 ± 9.74	199.1
4	225	135	39.49 ± 0.88	34.01	201.2 ± 9.52	202.8
5	150	26.4	26.39 ± 0.69	20.03	139.6 ± 2.86	143.7
6	150	153.6	35.24 ± 1.26	34.82	208.7 ± 10.00	203.8
7	44	90	18.93 ± 0.45	22.28	168.0 ± 8.44	168.1
8	256	90	39.86 ± 1.02	28.79	201.3 ± 14.54	205.0
9	150	90	33.38 ± 2.35	28.29	224.6 ± 12.50	224.3
10	150	90	32.54 ± 1.85	28.29	222.8 ± 11.08	224.3
11	150	90	29.40 ± 2.15	28.29	235.7 ± 7.06	224.3

Regarding the antioxidant activity, TEAC values between 139.6 and 235.7 µmol TE/g of protein were obtained. The highest TEAC values (>220 µmol ET/g protein) corresponded to the central points (9–11), reinforcing that moderate E/S conditions and intermediate reaction times favoured the release of antioxidant peptides of adequate size compared to high *DH* treatments (4, 6, and 8).These findings support that high *DH* did not guarantee greater antioxidant activity, as excessive hydrolysis time can lead to the breakdown of BPs into free AAs, thus losing their activity [63]. The antioxidant activity of protein hydrolysates in the ABTS assay is believed to be due to the exposure of specific AAs, such as phenylalanine, tryptophan, tyrosine, valine, and arginine, present in the generated peptides, which interact with the $ABTS^{\bullet+}$ radical, promoting its neutralization and stabilization [9,64].

The SDS-PAGE electrophoretic profile of the 11 runs is presented in Supplementary Figure S6. The hydrolysates corresponding to treatments 4, 6, and 8 showed a considerable decrease in the protein bands of the FA (10, 14, 20, and 24 kDa) to smaller peptides (<10 kDa), which correlated with the high *DH* achieved (35.24 to 39.86%). However, these treatments failed to gain maximum TEAC values due to prolonged protein hydrolysis, resulting in the release of very short peptides with low antioxidant activity or free AAs [65]. In contrast, treatments 1, 5, and 7 showed remanent FA bands (10–15 kDa), correlated with the low *DH* values achieved (<26%), producing a low amount of antioxidant peptides. In the case of treatments at central points, 9, 10, and 11, a greater intensity of protein bands <10 kDa was evident, related to moderate GH and greater antioxidant capacity (see Table 2).

In summary, a correlation was evident between the protein profile and the *DH* and TEAC results. These results highlight the importance of optimizing the E/S ratio and reaction time since short or excessive protein hydrolysis can limit the production of antioxidant peptides.

In this study, the mathematical models were developed based on the *DH* and antioxidant activity (TEAC) values. These models not only elucidated the relationship between experimental conditions, such as E/S and hydrolysis time, but also allowed the identification of significant interactions between these variables. The mathematical models representing *DH* and TEAC of the optimized HAP were expressed using the following equations:

$$DH(\%) = 30.53 + 5.42X_1 + 13.57X_2 \quad (3)$$

$$\text{TEAC} = 224.37 + 26.08X_1 + 39.9X_2 - 37.80X_1^2 - 48.82X_2^2 - 22.40X_1X_2 \quad (4)$$

The observed values closely matched those predicted by the model (Table 2), confirming the validity of the experimental approach and underscoring the importance of precise parameter control in designing functional processes. The analysis of variance (ANOVA) indicated that *DH* was best represented by a linear model based on its *p*-value, with the E/S ratio variable exhibiting a significant effect ($p = 0.0118$) (Table 3). In contrast, quadratic and interaction effects were not statistically significant ($p < 0.05$). This result suggested that increasing the E/S ratio had a positive effect on *DH* up to a stabilization point, consistent with findings from previous studies [66,67]. On the other hand, the $ABTS^{\bullet+}$ radical scavenging activity of the hydrolysates was best suited to a quadratic model, as indicated by a significant *p*-value ($p < 0.05$). This demonstrated the suitability of the model for optimizing hydrolysis conditions. Additionally, the "lack of fit" values were 0.759 and 0.647 for *DH* and TEAC, respectively, further confirming the adequacy of the models. The coefficient of determination (R^2), which measured the proportion of explained variation relative to total variation, approached unity for both models ($R^2 = 0.9618$ for *DH* and $R^2 = 0.9534$ for TEAC), indicating that these regression models accurately represented the system's behaviour. Similarly, the adjusted R^2 values (Adj $R^2 = 0.9363$ for *DH* and 0.9067 for TEAC) were high, supporting the model's significance.

Table 3. Analysis of variance (ANOVA) for the quadratic response surface model for the degree of hydrolysis (*DH%*) and ABTS$^{\bullet+}$ radical antioxidant activity (TEAC) of the optimized hydrolysate from the albumin fraction (F) of *Lupinus mutabilis* with protease PC7 (HAP).

Cause	Sum of Squares	Df	Middle Square	Value *f*	*p*
DH (%)					
(1) Time (L)	58.6758	1	58.6758	13.30566	0.067622
Time (Q)	1.3973	1	1.3973	0.31687	0.630181
(2) E/S (L)	368.5564	1	368.5564	83.57598	0.011755
E/S (Q)	8.2223	1	8.2223	1.86453	0.305398
Lack of Fit	8.4929	4	2.1232	0.48147	0.759349
Pure error	8.8197	2	4.4098		
R^2	0.9618				
R^2 adjusted	0.9363				
TEAC (μmol TE/g protein)					
(1) E/S (L)	1360.623	1	1360.623	11.97859	0.074298
E/S (Q)	2017.619	1	2017.619	17.76261	0.050195
(2) Time (L)	3187.316	1	3187.316	28.06033	0.033839
Time (Q)	3364.309	1	3364.309	29.61854	0.032144
1L by 2L	501.975	1	501.975	4.41927	0.170278
Lack of Fit	225.913	3	75.304	0.66296	0.647924
Pure error	227.176	2	113.588		
R^2	0.9534				
R^2 adjusted	0.9067				

The response surface methodology (RSM) provided a graphical representation of data analysis equations and was essential for assessing the influence and interactions among the factors. Figure 4A,B illustrate the contour plots and Figure 4C,D the three-dimensional surface graphs, revealing that the % of *DH* increased progressively toward optimal values, whereas antioxidant activity achieved its maximum value at an E/S ratio of 168 U/mg of substrate and a time of 105 min. To validate the accuracy of the model, a triplicate assay was conducted under the optimal conditions. The obtained values were 32.94 ± 0.53% for *DH* and 227.32 ± 23.78 μmol TE/g of protein for TEAC, showing no significant difference with the predicted values ($p < 0.05$). These findings confirmed that the response model was robust and suitable for predicting the optimization of the hydrolysis reaction.

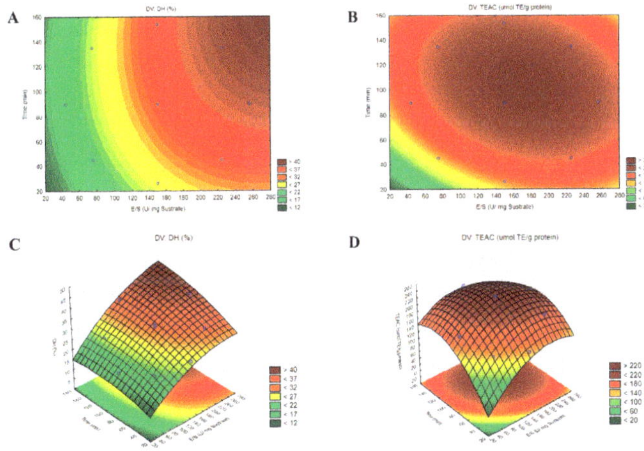

Figure 4. Contour plots (**A**,**B**) and three-dimensional surface plots (**C**,**D**) illustrating the effects of the enzyme-to-substrate ratio (E/S) and hydrolysis time (t) on the responses for the degree of hydrolysis (*DH%*) and ABTS$^{\bullet+}$ radical antioxidant activity (TEAC) in the optimized hydrolysate from the albumin fraction (AF) of *Lupinus mutabilis* with protease PC7 (HAP).

3.4. Hydrolysis of Albumin with Protease PC7 and Alcalase

Sequential hydrolysis was carried out using the PC7 protease and alcalase to enhance the biological activities of HAP [15]. The *DH* increased to 37.88 ± 0.51% after the action of alcalase (Figure 5A), a serine endopeptidase that cleaves peptide bonds at the C-terminal residues of AAs such as glutamic acid, methionine, leucine, tyrosine, lysine, and glutamine [68]. Various *DH* values have been reported for *L. mutabilis* hydrolysates. For example, Chirinos et al. achieved a *DH* of 33.11 ± 2.78% by hydrolysing the protein concentrate with alcalase at 0.385 AU/g protein for 90 min, while the combination with neutrase or flavourzyme did not result in a significant increase in *DH* [21]. Concerning the antioxidant capacity (TEAC value), a significant increase was evident, as shown in Table 4.

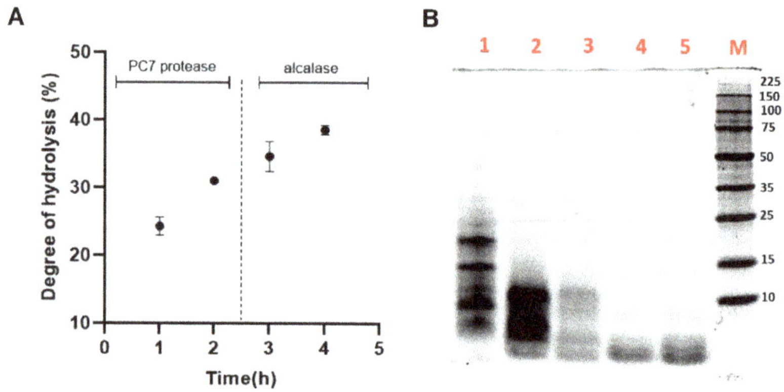

Figure 5. (**A**) Effect of sequential hydrolysis of the albumin fraction (AF) from *Lupinus mutabilis* albumin using protease PC7 and alcalase on the (**A**) degree of hydrolysis (*DH*); (**B**) protein profile analysed by SDS-PAGE under reducing conditions. Lane 1, non-hydrolysed AF; lanes 2 and 3, hydrolysis of AF with protease PC7 for 1 and 2 h, respectively; and lanes 4 and 5, hydrolysis of AF with alcalase for 1 and 2 h, respectively. M: molecular weight marker (Perfect Protein Marker 10–225 kDa, Sigma-Aldrich).

Table 4. Antioxidant activity (expressed as µmol TE/g for $ABTS^{\bullet+}$ and µmol TE/mg protein for ORAC), antihypertensive activity (expressed as IC_{50} for ACE inhibition), and antidiabetic activity (expressed in IC_{50} for DPP-IV inhibition) of the *Lupinus mutabilis* albumin fraction and its enzymatic hydrolysates (HAP and HAPA).

Sample	ABTS µmol TE/g Protein	ORAC µmol TE/g Protein	ACE IC_{50} (µg/mL)	DPP-IV IC_{50} (µg/mL)
Albumin fraction (AF)	92.0 ± 2.9 [a]	646.7 ± 55.3 [a]	179.6 ± 5.5 [a]	1207.9 ± 56.7 [a]
HAP	251.1 ± 11.0 [b]	1649.0 ± 75.7 [b]	24.9 ± 0.9 [b]	171.3 ± 10.3 [b]
HAPA	536.7 ± 6.8 [c]	3193.3 ± 230.3 [c]	13.5 ± 1.1 [c]	145.1 ± 6.8 [c]

[a,b,c] Different letters in the same column represent significant differences ($p < 0.05$).

Figure 5B illustrates the SDS-PAGE electrophoretic profile of the HAPA hydrolysate, providing a more detailed visualization of the peptide fractions generated during hydrolysis. The increase in *DH* and the reduction of the protein content of HAPA in comparison with HAP demonstrated that AF proteins were effectively degraded by both enzymes. Protein bands above 15 kDa were completely degraded after 60 min of hydrolysis with the PC7, while those below 15 kDa progressively decreased in intensity until the reaction's completion. The subsequent addition of alcalase facilitated the degradation of residual proteins in the HAP, ultimately producing bands smaller than 4 kDa, corresponding to the generated low MW and antioxidant peptides.

3.5. Assessment of Biological Activities

Results revealed that antioxidant activity varied depending on the assay used and increased progressively with hydrolysis time compared to the initial AF (Table 4). The optimized HAP exhibited a 2.7-fold and 2.5-fold increase in TEAC and ORAC values, respectively, compared to unhydrolysed AF. Sequential hydrolysis with PC7–alcalase (HAPA) further enhanced antioxidant activity, achieving a 2.1-fold increase compared to HAP. These findings underscore the beneficial impact of the hydrolysis process on the generation of peptides with potent antioxidant properties, aligning with the results reported by Intiquilla et al. [15]. Similarly, Sbroggio et al. observed a 45% increase in the antioxidant activity of Okara protein hydrolysate compared to the intact protein [69].

Previous studies indicate that the combination of alcalase with other enzymes can promote or decrease antioxidant activity [21,39]. In this study, the synergy between PC7 and alcalase significantly enhanced antioxidant activity, supporting the potential of this enzyme combination as a strategy for functional improvement. The antioxidant activity of protein hydrolysates can be attributed to the presence of low MW peptides enriched with aromatic (such as Tyr, Trp, Phe, and His), hydrophobic (such as Ala, Val, Leu, Pro, and Phe), and acidic (such as Glu and Asp) AAs that accounted for 27.64, 6.54, and 26.42% (Table 1) of the total AAs present, respectively [70,71]. Furthermore, the presence of leucine (3.73 g/100 g) and serine (3.63 g/100 g) in the released peptides has been shown to enhance antioxidant activity, as these amino acids are capable of donating electrons to stabilize the $ABTS^{\bullet+}$ radical [72]. Likewise, the presence of Glu + Glx (20.01 g/100 g) and Arg (8.54 g/100 g) favoured the solubility of these peptides and their antioxidant activity, due to their metal chelating capacity that catalyses the formation of reactive oxygen species (ROS) [73].

The multifunctionality of hydrolysates (HAP and HAPA) was further evaluated through their in vitro antihypertensive and antidiabetic activities by measuring their ACE and DPP-IV inhibitory properties (Supplementary Figure S7). Regarding the ACE inhibitory capacity, the AF showed an IC_{50} value of 179.6 μg/mL, which decreased significantly ($p < 0.05$) after hydrolysis with PC7 protease (24.9 μg/mL) and the combination PC7 + alcalase (13.5 μg/mL). These results surpassed those reported by Betancour et al., who reported an IC_{50} of 4.5 mg/mL for *Phaseolus vulgaris* protein hydrolysates with the alcalase–flavourzyme combination for 90 min [74]. The enzymatic digestion of *L. albus*, *L. angustifolius*, and *L. luteus* protein fractions with pepsin and trypsin yielded IC_{50} values between 200 and 300 μg/mL [75,76]. The IC_{50} values observed in this study were lower than that reported by Palma-Albino et al. for an AF from *Erythrina edulis* hydrolysed with a sequential combination of pepsin, pancreatin, and alcalase (50.65 μg/mL) [44]. These findings highlighted the variability in ACE inhibitory activity depending on the experimental conditions and the species studied, underscoring the need for optimizing enzymatic digestion processes to maximize the bioactivity of protein hydrolysates. Given the critical role of ACE in blood pressure regulation, particularly its contribution to hypertension, the identification and characterization of ACE inhibitory peptides represent a promising strategy for hypertension prevention.

In the case of the DPP-IV inhibitory activity, the IC_{50} value measured for the non-hydrolysed AF was 1207.9 μg/mL, which was significantly reduced ($p < 0.05$) to 171.3 μg/mL after being hydrolysed with PC7 protease and 145.1 μg/mL when both enzymes were used. The DPP-IV inhibitory activity observed in the present study was higher than that reported in previous research. Chirinos et al. obtained IC_{50} values between 2.14 and 3.21 mg/mL after hydrolysing the Tarwi protein concentrate with alcalase for 120 to 240 min [21]. These values were consistent with the IC_{50} value reported for the optimized protein hydrolysate of *L. albus* (3.42 mg/mL) obtained by sequential hydrolysis with alcalase, trypsin, and flavourzyme [77]. Similarly, the recent study by Muñoz et al., found that the gamma conglutin of *L. mutabilis* did not exhibit DPP-IV inhibitory activity.

However, after subjecting the fraction to gastric digestion, it managed to inhibit 100% of DPP-IV at a dose of 5 mg/mL [78].

This study demonstrates the potential of native proteases—used independently or in sequential hydrolysis with commercial enzymes—to produce multifunctional peptides with antioxidant, antihypertensive, and antidiabetic properties. By incorporating such protein hydrolysates into functional foods, the food industry can address critical health challenges, from chronic disease prevention to enhancing the techno-functional and sensory attributes of products. While soy, pea, hemp, and rice proteins dominate the market, rising consumer demand for innovative, health-promoting ingredients creates a compelling opportunity for *L. mutabilis*-derived hydrolysates to gain competitive traction. To fully unlock their commercial and nutritional potential, future research must prioritize the characterization of bioactive peptide fractions, including their precise identification, bioactivity validation, and bioaccessibility—key steps toward advancing functional food and nutraceutical development.

4. Conclusions

This study successfully produced a multifunctional protein hydrolysate from the AF of *L. mutabilis* seeds, obtained through TFF, demonstrating its potential as a high-value functional ingredient. The product obtained from the hydrolysis of AF with the PC7 protease under optimal conditions with an E/S ratio of 168 U/mg of substrate and a time of 105 min (HAP) showed high antioxidant activity, as determined by the ORAC method (1.649 μmol TE/g of protein). The multifunctionality of HAP was demonstrated after evaluating its ability to inhibit drug targets associated with hypertension (ACE) and diabetes (DPP-IV). Additionally, sequential hydrolysis of the PC7 protease, followed by alcalase, significantly improved its multifunctional properties (>1.5-fold). HAPA exhibited potent free radical scavenging ability in ABTS and ORAC assays, as well as strong inhibitory activities against ACE (IC_{50} = 13.5 μg/mL) and DPP-IV (IC_{50} = 145.1 μg/mL). These findings highlight the potential use of native PC7 proteases to produce multifunctional protein hydrolysates, potentially useful in the development of functional foods for the prevention of diseases associated with oxidative stress, hypertension, and diabetes. Finally, it is necessary to characterize the bioactive peptides produced in HAP and HAPA, as well as evaluate their bioavailability and bioactivity in biological systems.

Supplementary Materials: The following supporting information can be downloaded at: https://www.mdpi.com/article/10.3390/biotech14020032/s1, Figure S1: The complete OmicsBox workflow followed for the proteomic functional analysis of the albumin fraction of *Lupinus mutabilis* "Tarwi"; Figure S2: Chromatogram (amino acid analysis) of the albumin fraction of *Lupinus mutabilis* "Tarwi". Figure S3: Complete functional distribution of detected proteins in the cellular component functional group using gene ontology (GO).; Figure S4: Complete functional distribution of detected proteins in the biological process functional group using GO; Figure S5: Complete functional distribution of detected proteins in the molecular function functional group using GO; Figure S6: Electrophoretic profile (SDS-PAGE-Glycine) of the enzymatic hydrolysis optimization of the *Lupinus mutabilis* albumin fraction (AF) with PC7 protease; Figure S7: Evaluation of the bioactive properties of AF, HAP, and HAPA by the methods A. ABTS, B. ORAC, C. ACE inhibition, and D. DPP-IV inhibition at different sample concentrations; Table S1: Summary of key control parameters in the tangential ultrafiltration (TFF) process for concentrating albumin from *Lupinus mutabilis*.

Author Contributions: Performed the experiments, analysed the results, and wrote the manuscript, K.S.L.-B. and C.R.-V.; supervised and contributed to the design of the experiments, A.I. and A.I.Z.; reviewed and contributed to the statistical analysis of the optimization of enzymatic hydrolysis, M.S.; supervised and validated the determination of biological activities, K.J.-A.; performed the proteomic analysis, S.P.; reviewed, corrected, edited, and finalized the manuscript, B.H.-L. All authors have read and agreed to the published version of the manuscript.

Funding: This research was funded by Contract No. PE501078969-2022-PROCIENCIA, Peru, and Project COOPB23020 (CSIC, Spain).

Institutional Review Board Statement: Not applicable.

Informed Consent Statement: Not applicable.

Data Availability Statement: The original contributions presented in the study are included in the article/Supplementary Materials. Mass spectrometry proteomics data are available through ProteomeXchange under the identifier PXD062562; further inquiries can be directed to the corresponding authors.

Conflicts of Interest: The authors declare no conflicts of interest.

References

1. Rashidinejad, A. The Road Ahead for Functional Foods: Promising Opportunities amidst Industry Challenges. *Future Postharvest Food* **2024**, *1*, 266–273. [CrossRef]
2. Rizzello, C.G.; Tagliazucchi, D.; Babini, E.; Sefora Rutella, G.; Taneyo Saa, D.L.; Gianotti, A. Bioactive Peptides from Vegetable Food Matrices: Research Trends and Novel Biotechnologies for Synthesis and Recovery. *J. Funct. Foods* **2016**, *27*, 549–569. [CrossRef]
3. Singh, N.; Jain, P.; Ujinwal, M.; Langyan, S. Escalate Protein Plates from Legumes for Sustainable Human Nutrition. *Front. Nutr.* **2022**, *9*, 977986. [CrossRef] [PubMed]
4. Aderinola, T.A.; Duodu, K.G. Production, Health-promoting Properties and Characterization of Bioactive Peptides from Cereal and Legume Grains. *BioFactors* **2022**, *48*, 972–992. [CrossRef]
5. Shea, Z.; Ogando Do Granja, M.; Fletcher, E.B.; Zheng, Y.; Bewick, P.; Wang, Z.; Singer, W.M.; Zhang, B. A Review of Bioactive Compound Effects from Primary Legume Protein Sources in Human and Animal Health. *Curr. Issues Mol. Biol.* **2024**, *46*, 4203–4233. [CrossRef]
6. Chen, W.; Xu, D. Phytic Acid and Its Interactions in Food Components, Health Benefits, and Applications: A Comprehensive Review. *Trends. Food. Sci. Technol.* **2023**, *141*, 104201. [CrossRef]
7. Guerra-Ávila, P.L.; Guzmán, T.J.; Domínguez-Rosales, J.A.; García-López, P.M.; Cervantes-Garduño, A.B.; Wink, M.; Gurrola-Díaz, C.M. Combined Gamma Conglutin and Lupanine Treatment Exhibits In Vivo an Enhanced Antidiabetic Effect by Modulating the Liver Gene Expression Profile. *Pharmaceuticals* **2023**, *16*, 117. [CrossRef]
8. Hu, K.; Huang, H.; Li, H.; Wei, Y.; Yao, C. Legume-Derived Bioactive Peptides in Type 2 Diabetes: Opportunities and Challenges. *Nutrients* **2023**, *15*, 1096. [CrossRef]
9. Matemu, A.; Nakamura, S.; Katayama, S. Health Benefits of Antioxidative Peptides Derived from Legume Proteins with a High Amino Acid Score. *Antioxidants* **2021**, *10*, 316. [CrossRef]
10. Neji, C.; Semwal, J.; Kamani, M.H.; Máthé, E.; Sipos, P. Legume Protein Extracts: The Relevance of Physical Processing in the Context of Structural, Techno-Functional and Nutritional Aspects of Food Development. *Processes* **2022**, *10*, 2586. [CrossRef]
11. Zaky, A.A.; Simal-Gandara, J.; Eun, J.-B.; Shim, J.-H.; Abd El-Aty, A.M. Bioactivities, Applications, Safety, and Health Benefits of Bioactive Peptides from Food and By-Products: A Review. *Front. Nutr.* **2022**, *8*, 815640. [CrossRef] [PubMed]
12. López-García, G.; Dublan-García, O.; Arizmendi-Cotero, D.; Gómez Oliván, L.M. Antioxidant and Antimicrobial Peptides Derived from Food Proteins. *Molecules* **2022**, *27*, 1343. [CrossRef] [PubMed]
13. Fuel, M.; Mesas, C.; Martínez, R.; Ortiz, R.; Quiñonero, F.; Bermúdez, F.; Gutiérrez, N.; Torres, A.M.; Kapravelou, G.; Lozano, A.; et al. Antioxidant and Chemopreventive Activity of Protein Hydrolysates from Raw and Germinated Flour of Legumes with Commercial Interest in Colorectal Cancer. *Antioxidants* **2022**, *11*, 2421. [CrossRef]
14. Boschin, G.; Scigliuolo, G.M.; Resta, D.; Arnoldi, A. ACE-Inhibitory Activity of Enzymatic Protein Hydrolysates from Lupin and Other Legumes. *Food Chem.* **2014**, *145*, 34–40. [CrossRef]
15. Intiquilla, A.; Jiménez-Aliaga, K.; Zavaleta, A.I.; Hernández-Ledesma, B. Production of Antioxidant Hydrolyzates from a *Lupinus mutabilis* (Tarwi) Protein Concentrate with Alcalase: Optimization by Response Surface Methodology. *Nat. Prod. Commun.* **2018**, *13*, 751–756. [CrossRef]
16. Kamran, F.; Phillips, M.; Harman, D.G.; Reddy, N. Antioxidant Activities of Lupin (*Lupinus angustifolius*) Protein Hydrolysates and Their Potential for Nutraceutical and Functional Foods. *Food Chem. Adv.* **2023**, *2*, 100297. [CrossRef]
17. Lemus-Conejo, A.; Grao-Cruces, E.; Toscano, R.; Varela, L.M.; Claro, C.; Pedroche, J.; Millan, F.; Millan-Linares, M.C.; Montserrat-de La Paz, S. A Lupine (*Lupinus angustifolious* L.) Peptide Prevents Non-Alcoholic Fatty Liver Disease in High-Fat-Diet-Induced Obese Mice. *Food Funct.* **2020**, *11*, 2943–2952. [CrossRef]
18. Lermen, A.M.; Clerici, N.J.; Borchartt Maciel, D.; Daroit, D.J. Characterization and Application of a Crude Bacterial Protease to Produce Antioxidant Hydrolysates from Whey Protein. *Prep. Biochem. Biotechnol.* **2023**, *53*, 12–21. [CrossRef]

19. Bautista, C.; Arredondo-Nuñez, A.; Intiquilla, A.; Flores-Fernández, C.N.; Brandelli, A.; Jiménez-Aliaga, K.; Zavaleta, A.I. One-Step Purification and Characterization of a Haloprotease from *Micrococcus* Sp. PC7 for the Production of Protein Hydrolysates from Andean Legumes. *Arch. Microbiol.* **2024**, *206*, 377. [CrossRef]
20. Carvajal Larenas, F.E.; Linnemann, A.R.; Nout, M.J.R.; Koziol, M.; van Boekel, M.A.J.S. *Lupinus mutabilis*: Composition, Uses, Toxicology, and Debittering. *Crit. Rev. Food Sci. Nutr.* **2016**, *56*, 1454–1487. [CrossRef]
21. Chirinos, R.; Cerna, E.; Pedreschi, R.; Calsin, M.; Aguilar-Galvez, A.; Campos, D. Multifunctional in Vitro Bioactive Properties: Antioxidant, Antidiabetic, and Antihypertensive of Protein Hydrolyzates from Tarwi (*Lupinus mutabilis* Sweet) Obtained by Enzymatic Biotransformation. *Cereal Chem.* **2021**, *98*, 423–433. [CrossRef]
22. Ruiz-López, M.A.; Barrientos-Ramírez, L.; García-López, P.M.; Valdés-Miramontes, E.H.; Zamora-Natera, J.F.; Rodríguez-Macias, R.; Salcedo-Pérez, E.; Bañuelos-Pineda, J.; Vargas-Radillo, J.J. Nutritional and Bioactive Compounds in Mexican Lupin Beans Species: A Mini-Review. *Nutrients* **2019**, *11*, 1785. [CrossRef] [PubMed]
23. John, H.; Giri, S.K.; Subeesh, A.; Chandra, P.; Pandiselvam, R. Optimization of Process Parameters for the Production of Soy Protein by Ultrafiltration Using ANN. *J. Food Process. Preserv.* **2024**, *2024*, 5535413. [CrossRef]
24. Abril, A.G.; Pazos, M.; Villa, T.G.; Calo-Mata, P.; Barros-Velázquez, J.; Carrera, M. Proteomics Characterization of Food-Derived Bioactive Peptides with Anti-Allergic and Anti-Inflammatory Properties. *Nutrients* **2022**, *14*, 4400. [CrossRef]
25. Corrochano, A.R.; Cal, R.; Kennedy, K.; Wall, A.; Murphy, N.; Trajkovic, S.; O'Callaghan, S.; Adelfio, A.; Khaldi, N. Characterising the Efficacy and Bioavailability of Bioactive Peptides Identified for Attenuating Muscle Atrophy within a Vicia Faba-Derived Functional Ingredient. *Curr. Res. Food Sci.* **2021**, *4*, 224–232. [CrossRef]
26. Xu, R.; Sirtori, E.; Boschin, G.; Torres, K.B.; Arnoldi, A.; Aiello, G. Proteomic Analysis of the Seeds of Four Wild Mexican Lupinus Species: Focus on Storage Proteins. *Diversity* **2022**, *14*, 814. [CrossRef]
27. Dyall-Smith, M. *The Halohandbook*; University of Melbourne: Melbourne, Australia, 2009.
28. Sironi, E.; Sessa, F.; Duranti, M. A Simple Procedure of Lupin Seed Protein Fractionation for Selective Food Applications. *Eur. Food Res. Technol.* **2005**, *221*, 145–150. [CrossRef]
29. Bradstreet, R.B. Kjeldahl Method for Organic Nitrogen. *Anal. Chem.* **1954**, *26*, 185–187. [CrossRef]
30. AOAC International. *Official Methods of Analysis*, 22nd ed.; AOAC International: Rockville, MD, USA, 2023.
31. Torres-Sánchez, E.; Morato, E.; Hernández-Ledesma, B.; Gutiérrez, L.-F. Proteomic Analysis of the Major Alkali-Soluble Inca Peanut (*Plukenetia volubilis*) Proteins. *Foods* **2024**, *13*, 3275. [CrossRef]
32. Sanchiz, Á.; Morato, E.; Rastrojo, A.; Camacho, E.; González-de La Fuente, S.; Marina, A.; Aguado, B.; Requena, J.M. The Experimental Proteome of Leishmania Infantum Promastigote and Its Usefulness for Improving Gene Annotations. *Genes* **2020**, *11*, 1036. [CrossRef]
33. Shevchenko, A.; Wilm, M.; Vorm, O.; Mann, M. Mass Spectrometric Sequencing of Proteins from Silver-Stained Polyacrylamide Gels. *Anal. Chem.* **1996**, *68*, 850–858. [CrossRef]
34. Tran, N.H.; Qiao, R.; Xin, L.; Chen, X.; Liu, C.; Zhang, X.; Shan, B.; Ghodsi, A.; Li, M. Deep Learning Enables de Novo Peptide Sequencing from Data-Independent-Acquisition Mass Spectrometry. *Nat. Methods* **2019**, *16*, 63–66. [CrossRef]
35. Perez-Riverol, Y.; Bandla, C.; Kundu, D.J.; Kamatchinathan, S.; Bai, J.; Hewapathirana, S.; John, N.S.; Prakash, A.; Walzer, M.; Wang, S.; et al. The PRIDE Database at 20 Years: 2025 Update. *Nucleic Acids Res.* **2025**, *53*, D543–D553. [CrossRef] [PubMed]
36. Tran, N.H.; Zhang, X.; Xin, L.; Shan, B.; Li, M. De Novo Peptide Sequencing by Deep Learning. *Proc. Natl. Acad. Sci. USA* **2017**, *114*, 8247–8252. [CrossRef]
37. AltschuP, S.F.; Gish, W.; Miller, W.; Myers, E.W.; Lipman, D.J. Basic Local Alignment Search Tool. *J. Mol. Biol.* **1990**, *215*, 403–410. [CrossRef]
38. Nielsen, P.M.; Petersen, D.; Dambmann, C. Improved Method for Determining Food Protein Degree of Hydrolysis. *J. Food Sci.* **2001**, *66*, 642–646. [CrossRef]
39. Intiquilla, A.; Jiménez-Aliaga, K.; Iris Zavaleta, A.; Gamboa, A.; Caro, N.; Diaz, M.; Gotteland, M.; Abugoch, L.; Tapia, C. Nanoencapsulation of Antioxidant Peptides from *Lupinus mutabilis* in Chitosan Nanoparticles Obtained by Ionic Gelling and Spray Freeze Drying Intended for Colonic Delivery. *Food Biosci.* **2022**, *50*, 102055. [CrossRef]
40. Haider, S.R.; Reid, H.J.; Sharp, B.L. Tricine-SDS-PAGE. In *Electrophoretic Separation of Proteins*; Kurien, B.T., Scofield, R.H., Eds.; Methods in Molecular Biology; Springer: New York, NY, USA, 2019; Volume 1855, pp. 151–160. ISBN 978-1-4939-8792-4.
41. Re, R.; Pellegrini, N.; Proteggente, A.; Pannala, A.; Yang, M.; Rice-Evans, C. Antioxidant Activity Applying an Improved ABTS Radical Cation Decolorization Assay. *Free Radic. Biol. Med.* **1999**, *26*, 1231–1237. [CrossRef]
42. Hernández-Ledesma, B.; Dávalos, A.; Bartolomé, B.; Amigo, L. Preparation of Antioxidant Enzymatic Hydrolysates from α-Lactalbumin and β-Lactoglobulin. Identification of Active Peptides by HPLC-MS/MS. *J. Agric. Food Chem.* **2005**, *53*, 588–593. [CrossRef]
43. Hayakari, M.; Kondo, Y.; Izumi, H. A Rapid and Simple Spectrophotometric Assay of Angiotensin-Converting Enzyme. *Anal. Biochem.* **1978**, *84*, 361–369. [CrossRef]

44. Palma-Albino, C.; Intiquilla, A.; Jiménez-Aliaga, K.; Rodríguez-Arana, N.; Solano, E.; Flores, E.; Zavaleta, A.I.; Izaguirre, V.; Hernández-Ledesma, B. Albumin from Erythrina Edulis (Pajuro) as a Promising Source of Multifunctional Peptides. *Antioxidants* **2021**, *10*, 1722. [CrossRef]
45. Czubinski, J.; Feder, S. Lupin Seeds Storage Protein Composition and Their Interactions with Native Flavonoids. *J. Sci. Food Agric.* **2019**, *99*, 4011–4018. [CrossRef] [PubMed]
46. Asen, N.D.; Aluko, R.E.; Martynenko, A.; Utioh, A.; Bhowmik, P. Yellow Field Pea Protein (*Pisum sativum* L.): Extraction Technologies, Functionalities, and Applications. *Foods* **2023**, *12*, 3978. [CrossRef] [PubMed]
47. Loewe, D.; Grein, T.A.; Dieken, H.; Weidner, T.; Salzig, D.; Czermak, P. Tangential Flow Filtration for the Concentration of Oncolytic Measles Virus: The Influence of Filter Properties and the Cell Culture Medium. *Membranes* **2019**, *9*, 160. [CrossRef]
48. Hojilla-Evangelista, M.P.; Sessa, D.J.; Mohamed, A. Functional Properties of Soybean and Lupin Protein Concentrates Produced by Ultrafiltration-diafiltration. *J. Am. Oil Chem. Soc.* **2004**, *81*, 1153–1157. [CrossRef]
49. Nadal, P.; Canela, N.; Katakis, I.; O'Sullivan, C.K. Extraction, Isolation, and Characterization of Globulin Proteins from *Lupinus albus*. *J. Agric. Food Chem.* **2011**, *59*, 2752–2758. [CrossRef]
50. Muranyi, I.S.; Volke, D.; Hoffmann, R.; Eisner, P.; Herfellner, T.; Brunnbauer, M.; Koehler, P.; Schweiggert-Weisz, U. Protein Distribution in Lupin Protein Isolates from *Lupinus Angustifolius* L. Prepared by Various Isolation Techniques. *Food Chem.* **2016**, *207*, 6–15. [CrossRef]
51. Salmanowicz, B.P. Capillary Electrophoresis of Seed 2S Albumins from *Lupinus* Species. *J. Chromatogr. A* **2000**, *894*, 297–310. [CrossRef]
52. Salmanowicz, B.P.; Weder, J.K.P. Primary Structure of 2S Albumin from Seeds of *Lupinus albus*. *Eur. Food Res. Technol.* **1997**, *204*, 129–135. [CrossRef]
53. Foley, R.C.; Gao, L.-L.; Spriggs, A.; Soo, L.Y.; Goggin, D.E.; Smith, P.M.; Atkins, C.A.; Singh, K.B. Identification and Characterisation of Seed Storage Protein Transcripts from *Lupinus angustifolius*. *BMC Plant Biol.* **2011**, *11*, 59. [CrossRef]
54. Han, Y.; Ma, B.; Zhang, K. Spider: Software for Protein Identification from Sequence Tags with de Novo Sequencing Error. *J. Bioinform. Comput. Biol.* **2005**, *3*, 697–716. [CrossRef]
55. Rashid, M.H.U.; Yi, E.K.J.; Amin, N.D.M.; Ismail, M.N. An Empirical Analysis of Sacha Inchi (Plantae: *Plukenetia volubilis* L.) Seed Proteins and Their Applications in the Food and Biopharmaceutical Industries. *Appl. Biochem. Biotechnol.* **2024**, *196*, 4823–4836. [CrossRef]
56. Devkota, L.; Kyriakopoulou, K.; Bergia, R.; Dhital, S. Structural and Thermal Characterization of Protein Isolates from Australian Lupin Varieties as Affected by Processing Conditions. *Foods* **2023**, *12*, 908. [CrossRef]
57. Meyer, J.G. In silico Proteome Cleavage Reveals Iterative Digestion Strategy for High Sequence Coverage. *ISRN Comput. Biol.* **2014**, *2014*, 960902. [CrossRef] [PubMed]
58. Kruchinin, A.G.; Bolshakova, E.I.; Barkovskaya, I.A. Bioinformatic Modeling (In Silico) of Obtaining Bioactive Peptides from the Protein Matrix of Various Types of Milk Whey. *Fermentation* **2023**, *9*, 380. [CrossRef]
59. Salmanowicz, B.P. Primary Structure and Polymorphism of 2S Albumins from Seeds of Andean Lupin (*Lupinus mutabilis* Sweet). *Eur. Food Res. Technol.* **1999**, *209*, 416–422. [CrossRef]
60. Borek, S.; Pukacka, S.; Michalski, K. Regulation by Sucrose of Storage Compounds Breakdown in Germinating Seeds of Yellow Lupine (*Lupinus luteus* L.), White Lupine (*Lupinus albus* L.) and Andean Lupine (*Lupinus mutabilis* Sweet). II. Mobilization of Storage Lipid. *Acta Physiol. Plant.* **2012**, *34*, 1199–1206. [CrossRef]
61. Borek, S.; Pukacka, S.; Michalski, K.; Ratajczak, L. Lipid and Protein Accumulation in Developing Seeds of Three Lupine Species: *Lupinus luteus* L., *Lupinus albus* L., and *Lupinus mutabilis* Sweet. *J. Exp. Bot.* **2009**, *60*, 3453–3466. [CrossRef] [PubMed]
62. Islam, S.; Ma, W.; Appels, R.; Yan, G. Mass Spectrometric Fingerprints of Seed Protein for Defining *Lupinus* spp. Relationships. *Genet. Resour. Crop Evol.* **2013**, *60*, 939–952. [CrossRef]
63. Hunsakul, K.; Laokuldilok, T.; Sakdatorn, V.; Klangpetch, W.; Brennan, C.S.; Utama-ang, N. Optimization of Enzymatic Hydrolysis by Alcalase and Flavourzyme to Enhance the Antioxidant Properties of Jasmine Rice Bran Protein Hydrolysate. *Sci. Rep.* **2022**, *12*, 12582. [CrossRef]
64. Zheng, Y.; Wang, X.; Zhuang, Y.; Li, Y.; Tian, H.; Shi, P.; Li, G. Isolation of Novel ACE-Inhibitory and Antioxidant Peptides from Quinoa Bran Albumin Assisted with an In Silico Approach: Characterization, In Vivo Antihypertension, and Molecular Docking. *Molecules* **2019**, *24*, 4562. [CrossRef] [PubMed]
65. Krunic, T.; Rakin, M. Combination of Controlled Enzymatic Hydrolysis and Ultrafiltration as the Technique for Bioactive Peptide Production Intended for the Food Industry. *Innov. Food Sci. Emerg. Technol.* **2025**, *102*, 103979. [CrossRef]
66. Rostammiry, L.; Reza Saeidiasl, M.; Safar, R.; Javadian, R. Optimization of the Enzymatic Hydrolysis of Soy Protein Isolate by Alcalase and Trypsin. *Biosci. Biotechnol. Res. Asia* **2017**, *14*, 193–200. [CrossRef]
67. Sun, C.; Shan, Y.; Tang, X.; Han, D.; Wu, X.; Wu, H.; Hosseininezhad, M. Effects of Enzymatic Hydrolysis on Physicochemical Property and Antioxidant Activity of Mulberry (*Morus atropurpurea* Roxb.) Leaf Protein. *Food Sci. Nutr.* **2021**, *9*, 5379–5390. [CrossRef]

68. Tacias-Pascacio, V.G.; Morellon-Sterling, R.; Siar, E.-H.; Tavano, O.; Berenguer-Murcia, Á.; Fernandez-Lafuente, R. Use of Alcalase in the Production of Bioactive Peptides: A Review. *Int. J. Biol. Macromol.* **2020**, *165*, 2143–2196. [CrossRef]
69. Sbroggio, M.F.; Montilha, M.S.; Figueiredo, V.R.G.D.; Georgetti, S.R.; Kurozawa, L.E. Influence of the Degree of Hydrolysis and Type of Enzyme on Antioxidant Activity of Okara Protein Hydrolysates. *Food Sci. Technol.* **2016**, *36*, 375–381. [CrossRef]
70. Chen, C.; Sun-Waterhouse, D.; Zhang, Y.; Zhao, M.; Sun, W. The Chemistry behind the Antioxidant Actions of Soy Protein Isolate Hydrolysates in a Liposomal System: Their Performance in Aqueous Solutions and Liposomes. *Food Chem.* **2020**, *323*, 126789. [CrossRef]
71. Xiang, Z.; Xue, Q.; Gao, P.; Yu, H.; Wu, M.; Zhao, Z.; Li, Y.; Wang, S.; Zhang, J.; Dai, L. Antioxidant Peptides from Edible Aquatic Animals: Preparation Method, Mechanism of Action, and Structure-Activity Relationships. *Food Chem.* **2023**, *404*, 134701. [CrossRef]
72. Chen, T.; Chen, Z.; Wang, H.; Chen, X.; Yang, J.; Han, A.; Lin, D.-H.; Hong, J. Underlying Action Mechanism of a Novel Antioxidant Peptide Derived from Allium Tuberosum Rottler Protein Hydrolysates and Its Protective Effects on Hydrogen Peroxide Induced Cell Injury. *J. Funct. Foods* **2018**, *40*, 606–613. [CrossRef]
73. Wen, C.; Zhang, J.; Zhang, H.; Duan, Y.; Ma, H. Plant Protein-Derived Antioxidant Peptides: Isolation, Identification, Mechanism of Action and Application in Food Systems: A Review. *Trends Food Sci. Technol.* **2020**, *105*, 308–322. [CrossRef]
74. Betancur-Ancona, D.; Sosa-Espinoza, T.; Ruiz-Ruiz, J.; Segura-Campos, M.; Chel-Guerrero, L. Enzymatic Hydrolysis of Hard-to-Cook Bean (*Phaseolus vulgaris* L.) Protein Concentrates and Its Effects on Biological and Functional Properties. *Int. J. Food Sci. Technol.* **2014**, *49*, 2–8. [CrossRef]
75. Boschin, G.; Scigliuolo, G.M.; Resta, D.; Arnoldi, A. Optimization of the Enzymatic Hydrolysis of Lupin (*Lupinus*) Proteins for Producing ACE-Inhibitory Peptides. *J. Agric. Food Chem.* **2014**, *62*, 1846–1851. [CrossRef]
76. Yoshie-Stark, Y.; Bez, J.; Wada, Y.; Wäsche, A. Functional Properties, Lipoxygenase Activity, and Health Aspects of Lupinus Albus Protein Isolates. *J. Agric. Food Chem.* **2004**, *52*, 7681–7689. [CrossRef]
77. Rivero-Pino, F.; Leon, M.J.; Millan-Linares, M.C.; Montserrat-de La Paz, S. Antimicrobial Plant-Derived Peptides Obtained by Enzymatic Hydrolysis and Fermentation as Components to Improve Current Food Systems. *Trends Food Sci. Technol.* **2023**, *135*, 32–42. [CrossRef]
78. Muñoz, E.B.; Luna-Vital, D.A.; Fornasini, M.; Baldeón, M.E.; Gonzalez De Mejia, E. Gamma-Conglutin Peptides from Andean Lupin Legume (*Lupinus mutabilis* Sweet) Enhanced Glucose Uptake and Reduced Gluconeogenesis In Vitro. *J. Funct. Foods* **2018**, *45*, 339–347. [CrossRef]

BioTech

Review

Medicinal Honeys from Oceania: An Updated Review on Their Bioactive Constituents and Health Applications

Maryna Lutsenko [1,2], Michela Ravelli [1] and Gregorio Peron [1,*]

1 Department of Molecular and Translational Medicine, University of Brescia, 25123 Brescia, Italy; maryna.lutsenko@unibs.it (M.L.); ravellimichela5@gmail.com (M.R.)
2 Department of Professional Education, Restaurant and Tourism Business, Taras Shevchenko Luhansk National University, 36000 Poltava, Ukraine
* Correspondence: gregorio.peron@unibs.it

Abstract

Medicinal honeys from Oceania have gained considerable attention due to their peculiar bioactive constituents and potential health applications. Apart from small molecules such as methylglyoxal and hydrogen peroxide, these honeys are rich in phenolic compounds, volatile terpenes, and other bioactive molecules, which collectively contribute to their antioxidant, antimicrobial, anti-inflammatory, and wound-healing properties. Recent studies have highlighted the distinctive composition of Oceania honeys such as Manuka (*Leptospermum scoparium*), Jarrah (*Eucalyptus marginata*), and Agastache (*Agastache rugosa*) from New Zealand and Australia, demonstrating variability in bioactivity depending on floral source, geographical origin, and processing methods. This review synthesizes the current knowledge on the chemical profiles of these honeys with a particular focus on bioactive compounds and distinctive markers, and evaluates their therapeutic potential. Emphasis is placed on the mechanisms underlying their bioactivities, as well as emerging clinical and preclinical evidence supporting their medicinal use. By consolidating recent findings, this work provides an updated perspective on the functional properties of Oceania honeys, underscoring their relevance as natural products with significant health-promoting potential.

Keywords: honey; antimicrobial; antioxidant; Manuka; Jarrah; Agastache

Key Contribution: By adopting a comparative and mechanistic perspective, the work distinguishes methylglyoxal-driven non-peroxide activity in Manuka honey from peroxide- and polyphenol-mediated antimicrobial mechanisms in Jarrah and Agastache honeys, thereby clarifying similarities and differences in their therapeutic profiles. The review consolidates emerging evidence on scarcely studied Oceanian honeys, linking region-specific chemical signatures to biological function and highlighting their antimicrobial and antifungal potential.

1. Introduction

The term "honey" refers to the natural, sweet substance produced by *Apis mellifera* and other bee species from the nectar of flowers or the secretions of living plant parts, which they collect, transform, enrich with their own specific substances, and subsequently store and ripen in the comb [1,2]. As a globally traded commodity, its definition and quality standards are legally defined by regulatory bodies such as the Codex Alimentarius [3] and the EU Council Directive on honey [4]. Honey is broadly categorized by origin into floral (nectar-based) and honeydew (produced from insect secretions on plants) [5].

While historical accounts dating back over 8000 years highlight its traditional role as a food, ritual offering, and medicinal agent due to its antimicrobial and wound-healing properties [6,7], its modern value lies in its complex chemical composition and associated bioactive potential.

The botanical and geographical origin is the single most critical factor determining honey's composition, quality, and market value [8]. Verifying this origin is essential for controlling food fraud and guaranteeing product efficacy, driving the adoption of diverse methodological approaches, including the determination of sugar profiles, amino acid profiles, protein content, and phenolic compound profiles [9]. Traditionally, melissopalynological analysis (the study of pollen content) has been the primary method for origin determination, but it suffers from significant analytical limitations [10]. The pollen count is not exclusively related to the source nectar (primary enrichment) but is confounded by secondary enrichment (bee processing), tertiary enrichment (handling and extraction), and quaternary enrichment (airborne dispersion) [11]. Furthermore, factors such as hive migration and the need for highly specialized personnel and stringent method standardization render pollen analysis often inconclusive for definitive origin verification. This analytical ambiguity creates a critical gap in authentication practices.

To overcome these analytical limitations and provide a more robust basis for both authentication and therapeutic assessment, attention has shifted towards the non-volatile chemical fraction, specifically the polyphenols. Polyphenols are considered promising markers of honey authenticity because their content is significantly influenced by the botanical source, acting as a unique chemical fingerprint [12,13]. The application of High-Performance Liquid Chromatography combined with Mass Spectrometry (LC-MS) is now fundamental, allowing for the comprehensive separation and identification of complex phenolic profiles based on structure and mass [14].

In recent years, research has increasingly focused on honeys produced from distinctive botanical and geographical contexts, where unique floral sources give rise to specific chemical profiles and bioactive properties. Among these, honeys from Oceania including monofloral varieties from New Zealand and Australia have attracted considerable attention due to their particular composition in bioactive compounds and demonstrated antimicrobial, antioxidant, anti-inflammatory, and wound-healing activities. These findings highlight the potential of Oceania honeys as medicinal natural products and provide a strong rationale for their systematic evaluation.

The aim of this review is to provide a critical and structured overview of medicinal honeys from Oceania by integrating information on their chemical composition, bioactive constituents, and region-specific pharmacological evidence. Following a general examination of honey's chemical and nutritional properties, the review focuses on selected monofloral honeys from New Zealand and Australia, highlighting how botanical origin influences the presence of distinctive chemical markers and bioactive compounds. Particular attention is given to mechanistic insights and preclinical and clinical evidence supporting their therapeutic potential.

2. Literature Search and Selection Methodology

The literature included in this review was identified through a structured search strategy following general PRISMA guidelines for narrative reviews. Electronic searches were conducted in major scientific databases, including Scopus and PubMed, covering publications up to 2025. Additional relevant articles were identified by screening the reference lists of key reviews and original research papers. The search strategy combined keywords related to honey composition and bioactivity with terms specific to Oceania honeys. Search terms included combinations of "honey", "medicinal honey", "Manuka",

https://doi.org/10.3390/biotech15010005

"Jarrah", "Leptospermum", "Eucalyptus", "Agastache", "Oceania", "Australia", "New Zealand", "polyphenols", "chemical composition", "authentication", "antimicrobial", "antioxidant", "anti-inflammatory", and "wound healing". Boolean operators ("AND", "OR") were used to refine searches. Studies were included if they (1) reported original experimental, preclinical, or clinical data, or constituted relevant review articles; (2) focused on honeys specifically originating from Oceania; and (3) provided information on chemical composition, bioactive constituents, authentication markers, or biological activity. Exclusion criteria comprised studies not specifying botanical or geographical origin, articles lacking primary analytical or biological data, and non-peer-reviewed sources.

The review primarily considers studies published between 2019 and 2025. This period was selected because foundational studies characterizing honeys from botanical sources such as *Agastache* spp. and comprehensive comparative analyses including Jarrah honey were first published around 2019. Subsequent research has meaningfully expanded the field by providing new analytical insights, emerging composition-function relationships, and additional pharmacological evidence. Focusing on this recent period differentiates the present review from earlier general honey reviews and emphasizes recent progress, particularly for less-studied honeys such as Agastache and Jarrah.

3. Chemical and Nutritional Composition of Honey

Honey is a complex matrix composed of over 180 identified compounds, the relative concentrations of which are intrinsically determined by botanical and geographical origin, environmental conditions, bee species, and post-harvest processing methods [15,16]. This compositional variability dictates differences in its physical attributes, including color, viscosity, flavor, and bioactivity [15]. Honey typically consists primarily of carbohydrates (approximately 75–80%) and water (17–20%), with a minor, yet functionally essential, fraction (<5%) comprising non-sugar components such as proteins, enzymes, amino acids, polyphenols, and minerals (Table 1) [17,18].

Table 1. Average chemical composition of honey per 100 g.

Proximate	g/100 g	Minerals	mg/100 g	Vitamins	mg/100 g
Fructose	38.2	Potassium	40.0–3500.0	Ascorbic acid	2.2–2.5
Glucose	31.3	Calcium	3–31	Niacin	0.1–0.2
Water	17.1	Phosphorus	2.0–15.0	Pantothenic acid	0.02–0.11
Other disaccharides	5.0	Sodium	1.6 17.0	Pyridoxine	0.01–0.32
Sucrose	0.7	Magnesium	0.7–13.0	Riboflavin	0.01–0.02
Organic acids	0.5	Zinc	0.05–2.00	Thiamin	0.0–0.01
Proteins, amino acids	0.3	Iron	0.03–4.00		
		Manganese	0.02–2.0		
		Copper	0.02–0.60		
		Selenium	0.001–0.003		

3.1. Carbohydrates

The predominant constituents of honey are monosaccharides, specifically fructose (typically 38–41%) and glucose (31–35%), which collectively account for over 85% of the total sugar content [19]. These simple sugars are formed through the enzymatic hydrolysis of sucrose catalyzed by invertase (or sucrase) secreted by the bees during the trophallaxis process [20,21]. The ratio and concentration of these sugars significantly influence honey's primary physicochemical characteristics, including its water activity, crystallization rate, viscosity, and rheological behavior [22,23], but also its nutritional properties.

Minor carbohydrate fractions include disaccharides and oligosaccharides (typically <10%). While maltose and sucrose are the most common disaccharides, over a dozen

 https://doi.org/10.3390/biotech15010005

others (e.g., turanose, kojibiose, melibiose) have been identified. A significant proportion of oligosaccharides (trisaccharides, tetrasaccharides, and pentasaccharides) are neo-formed and are not directly derived from nectar. Their origin is attributed to trans-D-glycosylation reactions and non-enzymatic acid reversion (condensation of monosaccharides), both favored by the high sugar concentration and acidic pH of the honey matrix [24–29].

Given that carbohydrate profiles can vary significantly with botanical origin, their determination (commonly performed by HPLC, gas chromatography (GC), and MS [30]) is a robust tool. This analytical approach is critical for detecting adulteration, specifically the addition of exogenous syrups (e.g., corn, rice) which exhibit distinct sugar profiles compared to natural honey.

3.2. Phenolic Compounds

The non-sugar fraction of honey is fundamental to its bioactivity. Polyphenols, which are secondary plant metabolites, are primarily transferred to honey through flower nectar, pollen, and propolis [31]. These compounds significantly contribute to the honey's color, flavor, and, most importantly, its antioxidant capacity [13]. The two primary classes of polyphenols in honey are flavonoids and phenolic acids [31].

3.2.1. Flavonoids

Flavonoids represent the largest and most structurally diverse class of phenolic compounds, characterized by a C_6-C_3-C_6 flavane nucleus. Their inherent diversity arises from structural modifications such as hydroxylation, *O*-methylation, dimerization, and glycosylation reactions [32,33], which classify them into subclasses including flavonols, flavones, flavanones, and anthocyanidins [33,34]. In honey, the total flavonoid content can reach up to 6 mg/kg [35], with common examples including pinocembrin, apigenin, quercetin, galangin, and pinobanksin [13,36]. The specific qualitative and quantitative profile of flavonoids serves as a highly reliable chemical biomarker for determining the botanical origin of honey.

Therapeutic relevance of flavonoids stems from their well-documented anti-inflammatory, antimicrobial, antiallergic, antiviral, antioxidant, and cytotoxic activities. For instance, the 5,7-dihydroxyflavanone pinocembrin exerts anti-inflammatory, antifibrotic, and antioxidant activities [37–41], while quercetin and galangin exert a wide antimicrobial effect against bacteria, viruses and fungi of clinical interest, as demonstrated by several studies [42].

3.2.2. Phenolic Acids

These compounds, derived from either benzoic acid or cinnamic acid, are also crucial components originating from plant nectar [43–49]. They exist predominantly in ester, amide, or glycosidic forms [50] Common derivatives of benzoic acid include *p*-hydroxybenzoic, protocatechuic, and vanillic acids, while cinnamic acid derivatives include caffeic, *p*-coumaric, ferulic, and sinapic acids [51] The phenolic profile of honey depends on the botanical and geographical origin and contributes to its antioxidant capacity, together with flavonoids and enzymatic antioxidants. Common antioxidants in honey include gallic acid, caffeic acid, coumaric acid, and chlorogenic acid [13,48].

The antioxidant capacity of honey is significantly dependent on the presence of phenolic compounds. A 2020 study found a significant correlation between phenolic content, antioxidant capacity, and color. The total phenolic acid content of the honeys analyzed ranged from 3.57 to 18.83 mg CAE/100 g (concentration of phenolic acids in 100 g of honey), with the highest content observed in buckwheat honey and the lowest in clear honey [52].

https://doi.org/10.3390/biotech15010005

3.3. Minor Components: Organic Acids, Proteins, and Minerals

Organic acids (OAs) represent a small but critical fraction that profoundly influences honey's flavor and long-term stability, due to its resulting low pH (typically 3.2 to 4.5) [53]. Gluconic acid is the most predominant organic acid, formed via the oxidation of glucose by the enzyme glucose oxidase, a process integral to honey maturation. Other OAs, such as lactic, acetic, and citric acids, may originate from enzymatic synthesis by bees, plant secretions, or microbial fermentation [54–56].

Honey contains trace amounts of proteins and amino acids (e.g., proline, arginine), primarily derived from glandular secretions of the bees [19]. Key protein components are the enzymes, including: invertase (sucrase), which hydrolyzes sucrose into glucose and fructose; diastase (amylase), which breaks down starch; catalase and phosphatase. Enzyme activity, particularly diastase activity, is a globally recognized quality parameter used to verify that honey has not been subjected to excessive heat treatment (which denatures enzymes) [31].

Minerals, though typically less than 1% of the composition, are highly variable; darker honeys usually contain higher concentrations [16]. Potassium is the most abundant element (45–85% of total minerals), followed by calcium, sodium, magnesium, and trace elements like zinc, iron, and copper [16,57]. The mineral content not only impacts nutritional value but also serves as a partial geo-marker for origin determination.

Other minor compounds can be determined in honey as contaminants. The presence of heavy metals (e.g., cadmium, lead, arsenic) or pesticide residues acts as a direct bioindicator of environmental pollution in the forage area where the bees collected nectar [58]. Other compounds can be derived from processing. The most critical marker of poor handling or excessive heating is 5-hydroxymethylfurfural (HMF). HMF is a six-carbon heterocyclic compound that forms via the Maillard reaction from reducing sugars in acidic environments or through sugar self-degradation [57]. High HMF levels are universally regulated by quality standards (e.g., Codex Alimentarius, EU Directive) as an indicator of freshness and quality maintenance during processing and storage [59].

4. Therapeutic and Pharmacological Properties of Honey

The application of beehive products for therapeutic purposes, known as apitherapy, is experiencing a resurgence in evidence-based medicine [60]. Honey, in particular, exhibits a multifaceted spectrum of biological activities driven by its complex chemical matrix. The most compelling activities, explored below, include its potent antioxidant, antimicrobial, anti-inflammatory, and emerging antitumor potential.

4.1. Antioxidant Activity

Several components of honey such as ascorbic acid, carotenoids, and enzymes (e.g., glucose oxidase and catalase) have antioxidant properties and can act synergically. However, phenolic compounds have been reported as the main responsible of the antioxidant property [34]. The botanical and geographical origin is the paramount factor determining antioxidant capacity, with a strong correlation observed between the total content of phenolic acids and overall antioxidant activity [7]. Darker honeys are generally believed to contain higher concentrations of these phenolics, linking color to antioxidant potential [7]. A key advantage of honey is its lack of a pro-oxidant effect when administered therapeutically, a phenomenon sometimes observed with high doses of isolated classic antioxidants like vitamins C and E [61,62]. This protective effect is attributed to the complex interplay of multiple compounds present in the honey matrix. The systemic benefits of honey's antioxidant action are supported by preclinical models: in rats, pretreatment with wild honey protected against epinephrine-induced cardiac disorders and vasomotor

https://doi.org/10.3390/biotech15010005

dysfunction, highlighting its cardioprotective effects [63]. Tualang honey demonstrated a hypoglycemic effect in diabetic rats while improving renal oxidative stress [61]. The ester derivatives of caffeic acid (present at 20–25% in honey) are thought to contribute to its cytotoxic activity, with in vitro studies confirming significant inhibition of human hepatocellular carcinoma (HepG2) cell line proliferation [64].

4.2. Antimicrobial Activity

Honey's capacity to counteract various pathogenic organisms is attributed to several mechanisms, including its high sugar content and low pH (creating an inhospitable environment), the controlled enzymatic production of hydrogen peroxide, and the presence of non-peroxide components such as polyphenolic compounds and antimicrobial peptides [65,66].

The therapeutic use of honey is gaining traction, particularly for infections caused by antibiotic-resistant pathogens like Methicillin-Resistant *Staphylococcus aureus* (MRSA) [67,68]. Its broad-spectrum efficacy, demonstrated in vitro, positions it as a potential adjunct treatment to combat the rising threat of bacterial resistance [68].

Regarding honey's activity against fungi, in vitro efficacy has been reported against fluconazole-resistant fungal infections, such as those caused by *Candida albicans* in immunocompromised patients, suggesting a future role against drug-resistant fungal strains [31,69].

Finally, regarding antiviral activity, honey has demonstrated the ability to accelerate the healing time of skin lesions caused by HSV compared to standard antiviral agents like acyclovir [70].

4.3. Anti-Inflammatory and Wound Healing Activity

Honey's anti-inflammatory properties are linked to its bioactive constituents, notably flavonoids, which modulate the expression of pro- and anti-inflammatory cytokines (e.g., TNF-alpha, IL-2, IL-10, IL-12p70, and Interferon-gamma) [71].

Flavonoid extracts from honey have been shown to reduce the production of pro-inflammatory mediators in activated microglia, suggesting a potential role in attenuating the neuroinflammation associated with neurodegenerative diseases like Alzheimer's and Parkinson's [72]. Honey inhibits the generation of ROS and can modulate cyclooxygenase-2 activity [73]. Clinical administration of honey in healthy adults has been shown to reduce plasma concentrations of key inflammatory markers, including thromboxane B2, PGE2, and PGF2-alpha, supporting its use as an adjunct anti-inflammatory agent [73].

Honey is widely recognized in wound care. A seminal clinical study comparing honey to silver sulfadiazine for burn wounds found that honey treatment resulted in faster epithelialization, earlier resolution of acute inflammation, and improved infection control [74]. This efficacy is due to its physical barrier function combined with its antimicrobial and anti-inflammatory effects [75]. Evidence suggests also that honey may stimulate the proliferation and differentiation of stem cells, indicating a potential role in regenerative medicine [76,77].

4.4. Antitumor Activity

Honey exhibits promising antitumor properties by modulating critical cellular mechanisms, primarily apoptosis and cell proliferation [21,31]. In vitro studies show that honey depolarizes the mitochondrial membrane and activates the intrinsic apoptotic pathway by stimulating the expression of pro-apoptotic factors (e.g., p53 and caspases 3 and 9) while reducing anti-apoptotic factors (e.g., Bcl-2) [31]. Oral co-administration of honey and *Aloe vera* has been shown to decrease the nuclear protein Ki-67, a marker present during cell proliferation phases. This reduction slows the division rate of cancer cells, thereby inhibiting tumor growth [31].

https://doi.org/10.3390/biotech15010005

Honey can activate the immune system to counteract tumor growth, and specifically proteins such as apalbumin-1 and apalbumin-2 can induce the release of TNF-alpha from macrophages [78–81]. The molecular mechanisms underlying these effects include the modulation of oxidative stress, reduction in inflammation, and inhibition of angiogenesis in tumor cells [21]. While preclinical evidence is compelling, scientific data on honey's efficacy as a primary antitumor agent is currently limited to cellular studies, necessitating further investigation in more complex in vivo and clinical models to fully confirm its therapeutic potential [31]. The path forward requires targeted research to link the chemical composition of specific therapeutic honeys to their defined pharmacological outcomes.

5. Chemical and Pharmacological Properties of Oceanian Medicinal Honeys

This section shifts the focus to an in-depth analysis of high-value medicinal honeys from the Oceania region, specifically Manuka, Agastache, and Jarrah honeys. The distinct therapeutic profiles of these honeys are fundamentally driven by unique, non-peroxide chemical components.

5.1. Manuka Honey

Manuka honey, derived from the nectar of *Leptospermum scoparium* (Myrtaceae), is a dark, monofloral honey primarily produced in New Zealand and Australia [82–87]. It is globally recognized for its unique characteristics, including a distinctive earthy flavor, variable color (deep cream to amber), and a robust, non-peroxide-based antibacterial activity [88–90].

The phytotherapeutic use of Manuka honey is largely attributed to methylglyoxal (MGO), a reactive alpha-oxoaldehyde [91]. MGO is spontaneously formed during honey maturation from its precursor, dihydroxyacetone (DHA), which accumulates in high concentrations specifically in the nectar of *L. scoparium* [92]. The formation of MGO is catalyzed by the unique physiological environment of the honey: DHA, present in the nectar, is converted to MGO during the maturation period, driven by the actions of the bees and time [93–96]. The conversion rate is influenced by factors such as storage conditions and moderate heat, which can be intentionally utilized post-harvest to increase MGO concentration, thereby enhancing the antibacterial efficacy [97]. The concentration of MGO in Manuka honey typically ranges significantly, from 30 to over 800 mg/kg, and is linearly correlated with its unique and potent antibacterial activity [91]. While MGO is not exclusive to Manuka honey (being detected at substantially lower concentrations in other varieties, such as 0.4–17 mg/kg in some Italian honeys [98,99]), its high concentration remains the definitive chemical marker.

The quality and efficacy of Manuka honey are standardized using the Unique Manuka Factor (UMF), a standardized scoring system developed in New Zealand [91]. The UMF is a non-physical numerical index that primarily reflects the concentration of MGO (and its precursor DHA), in addition to other compounds like leptosperin [91]. Higher UMF ratings indicate a stronger overall antibacterial potency, effective against both Gram-positive and Gram-negative pathogens [91]. To ensure product authenticity and distinguish Manuka from botanically similar honeys (e.g., Kanuka, Kunzea ericoides), specific chemical markers are utilized alongside MGO, including leptosperin, lepteridine, 3-phenyllactic acid, 4-hydroxyphenyllactic acid, 2-methoxybenzoic acid (*o*-anisic acid), and 2′-methoxyacetophenone (Figure 1). Some of these compounds have been detected also in other medicinal honeys, although in lower amounts and different ratios. Leptosperin has been specifically highlighted as a unique chemical marker identified in Manuka honey. These compounds have also been associated with the antimicrobial properties of Manuka

 https://doi.org/10.3390/biotech15010005

honey [90]. Furthermore, advanced palynological techniques allow for the differentiation of *L. scoparium* pollen from *K. ericoides* based on morphological distinctions like equatorial diameter and surface texture [100].

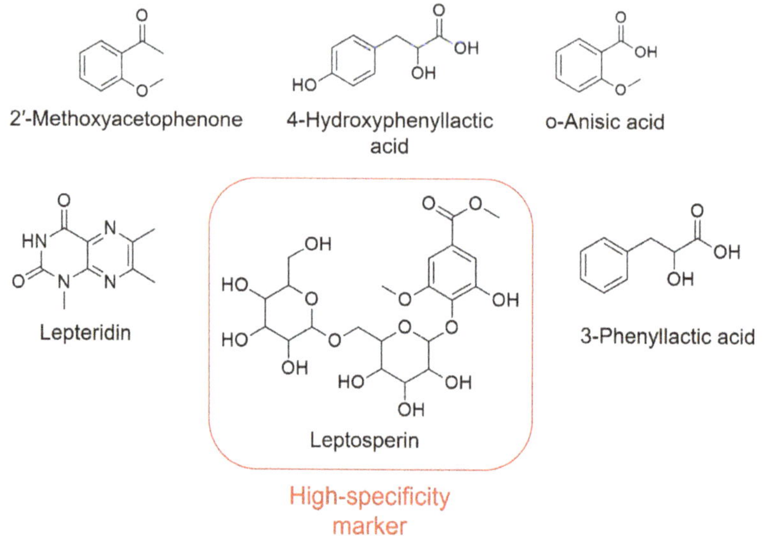

Figure 1. Chemical markers of Manuka honey.

In addition to these constituents, phenolic acids such as protocatechuic acid, syringic acid, and gallic acid are present, as well as flavonoids such as quercetin, luteolin, apigenin, kaempferol, galangin, pinobanksin, and pinocembrin. The total phenolic content (TPC) of manuka honey samples ranges from 250 to 985 μg/g, depending on the geographic origin and storage and processing conditions. Among the identified phenolic acids, gallic acid represents approximately 49% of the total phenolic acids, followed by syringic acid which represents 44%. Among the flavonoids, quercetin represents 46% of the total content, luteolin 33% and kaempferol 15%.

Among the vitamins, the most represented are thiamine (B1), riboflavin (B2), ascorbic acid (vitamin C), pantothenic acid and pyridoxine. Vitamin C in this type of honey has a concentration equal to 1067.4 mg/kg. All these phytocompounds are responsible for the antimicrobial, antioxidant, anti-inflammatory, anti-tumor, immunomodulatory and wound healing properties of manuka honey [89]. Based on this distinctive chemical composition, Table 2 summarizes representative in vitro data on the antibacterial activity of Manuka honey against clinically relevant microorganisms.

Table 2. Exemplificative data showing the antibacterial activity of Manuka honey.

Targeted Microbe	Efficacy	Active Constituents	Ref.
Staphylococcus aureus	MIC: 1.1 mM (MGO) MIC: 4.3 mM (GO)	MGO/GO	[82]
Escherichia coli	MIC: 1.1 mM (MGO) MIC: 6.9 mM (GO)	MGO/GO	[82]
Pseudomonas aeruginosa	MIC: 12% *w*/*v* MBC: 16% *w*/*v*	N/R	[83]
Salmonella enterica	MIC = 8.9% (*v*/*v*)	N/R	[84]

https://doi.org/10.3390/biotech15010005

Table 2. *Cont.*

Targeted Microbe	Efficacy	Active Constituents	Ref.
Pseudomonas aeruginosa (ATCC 27853)	MIC = 9.5% (*w*/*v*); MBC = 12% (*w*/*v*)	N/R	[91]
Pseudomonas aeruginosa (biofilm)	MIC: 5–7% *w*/*v*; MBC: 15% *w*/*v*	N/R	[92,93]
E. coli, Salmonella typhimurium, Salmonella enteritidis, Salmonella mississippi, Yersinia enterocolitica, Enterobacter aerogenes, Enterobacter cloacae, Shigella flexneri, Shigella sonnei	MIC: 5–10% (*v*/*v*); MBC: 5–10% (*v*/*v*); *Enterobacter* spp. MIC/MBC = 10–17%	H_2O_2, MGO	[93]
Staphylococcus aureus	MIC_{50}: UMF 5+ = 6%, UMF 10+ = 7%, UMF 15+ = 15%; for Enterobacteriaceae, MIC_{50} ~21–27%	MGO	[94]
Bacillus subtilis	MIC = 0.8 mM	MGO	[95]
Escherichia coli	MIC = 1.2 mM	MGO	[95]

UMF: Unique Manuka Factor: based on the measured level of MGO. UMF 5+: ≥83 mg/kg, UMF 10+: ≥263 mg/kg, UMF 15+: ≥514 mg/kg. MGO: methylglyoxal.

Consistent with this distinctive chemical profile, Manuka honey exhibits potent antimicrobial activity, which has elevated its status from a "superfood" [101] to an approved Medical-Grade Honey (MGH) used clinically, particularly in advanced wound and burn care. MGH is subject to stringent quality and safety standards for medical use [91]. In compliance with international regulations (e.g., European Directive 2001, updated 2019, and Regulation (EU) 2017/745), MGH must be sterile, organic, and free of contaminants [101–103]. Sterility is typically achieved through gamma irradiation [102]. While this process may cause a minor reduction in antimicrobial potency, it is essential for ensuring microbiological safety in the clinical setting, allowing MGH to be governed under medical device regulations due to its proven wound management applications.

The antimicrobial effect is quantitatively assessed using the Minimum Inhibitory Concentration (MIC), defined as the lowest concentration of the honey or its pure compound (e.g., MGO) required to inhibit visible microbial growth [91]. For pure MGO, MIC values of 0.15–0.31 mg/mL have been reported against common pathogens such as *S. aureus* and *P. aeruginosa*.

5.1.1. Optimizing Bioactive Content: Correlation Between Flower Development and Metabolite Production

Optimizing manuka honey production requires maximizing nectar sugar levels while preserving high concentrations of DHA. This can be achieved through increased floral abundance or enhanced nectar volume, both of which are influenced by environmental factors such as leaf area and temperature, which govern sugar biosynthesis. Smallfield (2018) [90] classified eight distinct floral developmental stages (FDS), ranging from immature buds to post-anthesis flowers with wrinkled stamens and nectar-rich hypanthia. Total sugars and DHA progressively increased across development, peaking immediately before floral senescence. Authenticity markers of manuka honey, including leptosperin, lepteridine, and 3-phenyllactic acid, exhibited similar accumulation patterns, indicating biosynthetic linkage with nectar production [90].

Metabolite-specific trends were also observed: 2′-methoxyacetophenone reached maximal levels at stage 5, whereas 2-methoxybenzoic acid increased continuously throughout anthesis. In contrast, leptosperin and lepteridine peaked earlier than methylsyringate and 2-methoxybenzoic acid [90]. These findings demonstrate that the floral development stage is a critical determinant of nectar composition and, consequently, honey quality.

https://doi.org/10.3390/biotech15010005

5.1.2. Antimicrobial and Wound Healing Properties of Manuka Honey

Manuka honey has significant antibacterial activity [103]. The efficacy against *S. aureus* has been evaluated in comparison with sidr honey (SH) and tualang honey (TH) using agar diffusion, MIC, minimum bactericidal concentration (MBC), time-kill curve, microtiter plate, and RT-qPCR assays. All honeys demonstrated inhibitory activity, with Manuka exhibiting the largest inhibition zones [104]. MIC and MBC values confirmed greater potency for Manuka (12.5% and 25% v/v, respectively) compared to SH and TH (MIC: 25%, MBC: 50%). Biofilm formation was significantly reduced at 60% honey concentration, with Manuka achieving the highest inhibition (52%), followed by SH (46%) and TH (42%). Time-kill curves revealed reduced colony-forming units after honey exposure, while RT-qPCR demonstrated downregulation of multiple *S. aureus* virulence genes, with Manuka showing the strongest effects [104].

In the clinical context of wound management, the topical application of honey is beneficial for maintaining a moist healing environment, providing a protective barrier against exogenous and endogenous microbial contamination, and accelerating tissue repair [104]. This therapeutic action is enhanced by the relatively low pH (3.5–4.5) of Manuka honey, which contributes to antibacterial activity, promotes fibroblast activation and tissue oxygenation, and aids in mitigating tissue damage associated with ROS by limiting inflammation [103]. Antimicrobial efficacy can also be leveraged synergistically, as studies have reported positive Fractional Inhibitory Concentration Index (FICI) values when Manuka honey is combined with conventional antibiotics like tetracycline and oxacillin [105,106].

Despite its compelling clinical efficacy, the inherent viscosity and limited mechanical stability of raw honey present challenges for consistent topical dosing and application [107]. To address this, biomedical engineering approaches have focused on incorporating Manuka honey into advanced scaffold and hydrogel systems. For instance, researchers have successfully developed three-dimensionally printed honey-gelatin hydrogel patches that not only retained antibacterial properties but also actively stimulated fibroblast and keratinocyte proliferation and promoted angiogenesis [107]. Similarly, composite systems, such as 2-hydroxyethyl methacrylate/gelatin (HG) hydrogel scaffolds, have improved structural integrity, swelling behavior, and biocompatibility, utilizing gelatin as a versatile natural scaffold material [108]. These formulation efforts were validated by a comprehensive study that investigated the preparation of topical agents derived from various Western Australian honeys (including two Manuka varieties, Jarrah, and Coastal Peppermint) compared to New Zealand Manuka [109]. The research involved transforming the honeys into standardized honey-loaded sodium alginate pre-gel solutions and subsequent wet and freeze-dried sheets to overcome dosing inconsistencies [109]. Critically, the study found that the formulation process, including the use of sodium alginate and calcium chloride, did not compromise the antioxidant activity of the honeys, which was consistently measured via FRAP and HPTLC assays across pure, pre-gel, wet, and dry sheet forms [109]. Furthermore, Australian and New Zealand Manuka, along with Jarrah honey, demonstrated consistent growth inhibition zones against all tested Gram-positive and Gram-negative bacteria. Although pure honeys generally exhibited the highest activity, the dry sheets followed closely, confirming that formulations intended for in vivo application retain the crucial antioxidant and antibacterial activities, thereby supporting their suitability as standardized topical wound healing agents [109].

5.1.3. Quality Parameters of Manuka Honey

To ensure the high standard, safety, and authenticity required for a premium medicinal product, the quality of Manuka honey is governed by stringent international and national regulations. These regulations specify key physicochemical parameters, including mois-

https://doi.org/10.3390/biotech15010005

ture content, sugar composition, acidity, and electrical conductivity, which aim to protect consumers from deceptive practices and guarantee product integrity [97]. Two physicochemical indicators are particularly critical for quality control: diastase activity and HMF content [97]. As reported above, diastase is a key enzyme in honey, and a low level of its activity is considered an indicator of either excessive heat treatment or industrial processing, compromising the honey's freshness [97]. Conversely, HMF, a furan compound formed from sugar degradation under acidic conditions or high temperatures, is a critical marker of quality degradation and is also used to assess potential adulteration with processed sugar additives [97]. Beyond these general quality markers, the New Zealand Ministry of Primary Industries has established a specific five-characteristic profile necessary for authenticating Manuka honey and distinguishing monofloral from multifloral varieties [97]. Four of these characteristics are specific chemical compounds, i.e., 2′-methoxyacetophenone, 2-methoxybenzoic acid, 3-phenyllactic acid, and 4-hydroxyphenyllactic acid, while the fifth involves the use of DNA markers found in Manuka pollen, providing an authoritative, multi-modal standard for authenticity [97].

5.1.4. Methylglyoxal: Dual Role and Safety Considerations

The definitive component responsible for the potent, non-peroxide antibacterial activity of Manuka honey is MGO. However, MGO also presents a critical paradox, as it is a highly reactive alpha-oxoaldehyde and a potent glycotoxin widely implicated in the pathogenesis of chronic human diseases, including diabetes mellitus, neurodegenerative disorders (e.g., Alzheimer's and Parkinson's), and cardiovascular conditions (e.g., atherosclerosis and hypertension) [89,98]. MGO acts as a key precursor in the non-enzymatic reaction known as glycation, leading to the formation of Advanced Glycation End-products (AGEs) [89,98]. Glycation involves the reaction between the carbonyl groups of reducing sugars or dicarbonyl compounds (like MGO) and the amino groups of proteins or lipids [110]. This reaction, which proceeds through pathways such as the Maillard reaction (including the Hodge and Namiki pathways) [111,112], as well as the polyol and lipid peroxidation pathways, is a normal metabolic byproduct [111]. However, under conditions of metabolic stress or aging, the accumulation of high levels of AGEs in tissues and bloodstream induces oxidative stress and the generation of ROS [111]. Physiologically, cells possess sophisticated defense mechanisms, notably the glyoxalase system (comprising glyoxalase 1 and glyoxalase 2), which detoxifies MGO [113]. Nevertheless, when MGO synthesis exceeds this detoxification capacity, AGE accumulation occurs [113]. The subsequent binding of AGEs to their transmembrane receptors, known as Receptors for AGEs (RAGEs), initiates a cascade of intracellular events that activate inflammatory signaling pathways (e.g., NF-kB), leading to increased ROS production, inflammation, and cellular apoptosis [111].

Crucially, the potential toxicological effects of high MGO intake from Manuka honey must be balanced against the protective compounds inherently present within the honey matrix. The naturally abundant polyphenols and other antioxidants (e.g., ascorbic acid, glutathione) in honey are known to mitigate AGE damage [114]. These phenolic antioxidants exert protection through multiple mechanisms: (1) ROS scavenging: they are transformed into stable phenoxy radicals, directly attenuating oxidative stress induced by AGEs [114]; (2) chelation: their structure allows them to chelate metal ions (such as copper and iron), thereby reducing the formation of ROS [114]; (3) metabolic modulation: they can promote glucose metabolism by modulating key enzymes, such as AMP-activated protein kinase (AMPK), and activating insulin receptors [114].

This intrinsic chemical buffering capacity suggests that the consumption of Manuka honey, unlike exposure to isolated MGO, provides simultaneous administration of the

 https://doi.org/10.3390/biotech15010005

glycotoxin and a powerful complex of protective and detoxifying agents, underscoring the importance of consuming the entire honey matrix rather than isolated components.

Manuka honey has attracted considerable academic and commercial interest, while research data on other honeys is much sparser. Below, the chemical composition and the bioactive properties of Jarrah and Agastache honeys, two other species that have been investigated in recent years, are discussed.

5.2. *Jarrah Honey*

Jarrah honey is a monofloral honey derived exclusively from the nectar of the *Eucalyptus marginata* tree, a towering eucalypt species endemic to the geographically isolated south-west region of Western Australia [115]. The premium status of Jarrah honey is inherently linked to the specific biology of its source, as *E. marginata* exhibits an unusually long and sporadic flowering cycle. This tree produces white or cream-colored flowers only every three to four years, a frequency that is dependent on climatic conditions [115]. This infrequent and geographically specific spring flowering, typically occurring from September to January, causes the scarcity of the resulting honey, establishing it as a highly valued medicinal commodity.

From a physicochemical perspective, Jarrah honey is characterized by a low tendency to crystallize, a property primarily attributed to its high fructose-to-glucose ratio and generally low pollen content. Reduced pollen content limits the availability of nucleation sites that normally initiate crystallization, while the elevated fructose fraction (approximately 42.5%) further stabilizes the liquid state [110]. This compositional feature is also associated with a comparatively lower glycemic index (GI) relative to many other honey varieties, which has potential nutritional and therapeutic implications [11].

5.2.1. Phytochemical Profile and Chemical Markers

In recent years, analytical techniques have been crucial for defining the unique chemical signature of Jarrah honey. A 2023 study by Lawag and colleagues used high-performance thin-layer chromatography (HPTLC) to examine the phytochemical composition and antioxidant activity of Jarrah honey [116]. Compounds identified and regarded as chemical markers using this technique included hesperitin, *o*-anisic acid, taxifolin, kojic acid, *m*-coumaric acid, lumichrome, epigallocatechin gallate, and 2,3,4-trihydroxybenzoic acid. The authors sought to determine which honey constituents contribute to antioxidant activity using a novel analytical method based on hue changes upon contact with the reagent 2,2-diphenyl-1-picrylhydrazyl (DPPH) [116]. The reaction of compounds with DPPH is primarily governed by the steric accessibility of the reagent, meaning that smaller molecules have greater access to the radical site than larger molecules [116]. This explains why larger molecules like flavonoids tend to react more slowly than smaller molecules, such as simple phenolic acids [116]. The reactivity of flavonoids with DPPH is further dictated by the so-called Bors criteria, which outline specific structural requirements for enhanced antioxidant power [116]. While the HPTLC-DPPH assay is a powerful tool for identifying antioxidant constituents in vitro [116], its biological relevance is limited, as the DPPH radical does not exist naturally in biological or food systems. Therefore, the authors strongly suggest utilizing more biochemically relevant antioxidant models in future studies [116].

A separate study by Anand and colleagues focused on identifying the major volatile and semi-volatile bioactive compounds [117]. Their analysis showed that the dominant bioactive compounds in Jarrah honey are isophorone (40.06%), nonanoic acid (5.62%), 2-hydroxy-3,5,5-trimethyl-2-cyclohexenone (2-hydroxyisophorone; 5.12%), and 2,4-diisocyanato-1-methylbenzene (3.75%) (Figure 2) [117]. Isophorone was specifically

 https://doi.org/10.3390/biotech15010005

highlighted as a unique chemical marker identified only in Jarrah honey, establishing it as a key signature compound derived from the *E. marginata* source. Nonanoic acid was also identified as a major compound, though it was additionally found in Jelly bush and Tea tree honeys, suggesting its utility as a general marker of certain floral types rather than a specific Jarrah unique marker [117].

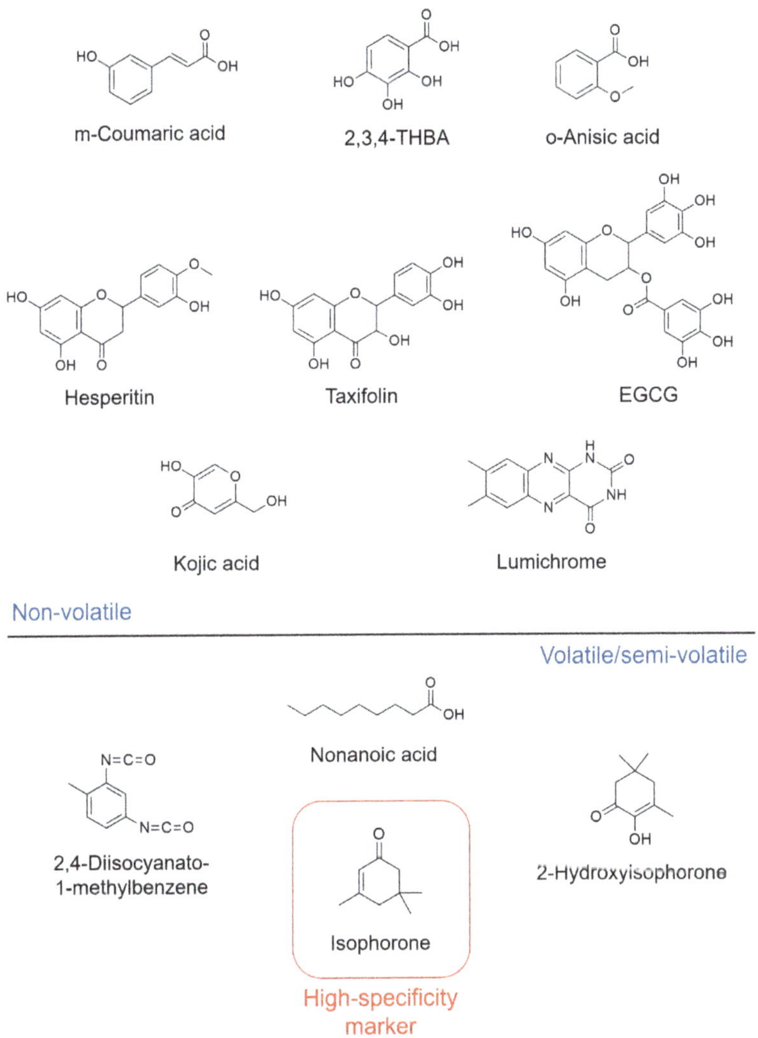

Figure 2. Chemical markers of Jarrah honey. 2,3,4,-THBA: 2,3,4-trihydroxybenzoic acid; EGCG: epigallocatechin gallate.

5.2.2. Antimicrobial Properties and Mechanism

On the basis of this distinctive chemical composition, Jarrah honey exhibits significant antimicrobial and antifungal activities. Its antibacterial efficacy, summarized in Table 3, is primarily associated with hydrogen peroxide (H_2O_2) production via the enzymatic activity of glucose oxidase, in combination with its rich phenolic profile [115]. Unlike Manuka honey, whose antimicrobial activity is largely driven by high concentrations of MGO, Jarrah honey represents a peroxide-dominant system, highlighting an important mechanistic distinction between these two medicinal honeys.

 https://doi.org/10.3390/biotech15010005

Table 3. Exemplificative data showing the antibacterial activity of Jarrah honey.

Targeted Microbe	Efficacy	Active Constituents	Ref.
Staphylococcus aureus	MIC: 4–12% *w*/*v*; MIC$_{50}$: 4–8% *w*/*v*	H_2O_2	[118]
Methicillin-resistant *Staphylococcus aureus* (MRSA)	MIC: 6–12% *w*/*v*	H_2O_2	[118]
Coagulase-negative staphylococci	MIC$_{50}$: 4–8% *w*/*v*	H_2O_2	[118]
Streptococcus pyogenes	MIC: 8–29% *w*/*v*	N/R	[119]
Enterococcus faecalis	MIC: 10–25% *w*/*v*	N/R	[119]
Escherichia coli	MIC: 5% *w*/*v*	N/R	[119]
Pseudomonas aeruginosa	MIC: 5–16% *w*/*v*	N/R	[82]
Staphylococcus epidermidis	MIC: 4–12% *w*/*v*	N/R	[120]
Candida albicans (ATCC 10231)	MIC: 20% *w*/*v*	N/R	[120]
Dermatophyte fungi (*Trichophyton rubrum, T. mentagrophytes*)	MIC: 1.5–3.5% *w*/*v*	H_2O_2	[121]
Microsporum canis	MIC: 1.5–3.5% *w*/*v*	H_2O_2	[121]
Other Gram-negative clinical isolates (general panel)	MIC: 6.7–28.0%	H_2O_2	[121]

The ranges of MIC are related to different Jarrah samples tested within the same study. N/R: not reported.

As with its antibacterial activity, while the production of H_2O_2 is essential, the level present in the honey appears too low to solely justify its potent antifungal activity [121]. Reported concentrations of H_2O_2 range from 0.04 to 4 mM, with an average value of approximately 2.5 mM. While enzymatic oxidation of glucose is the principal pathway for peroxide generation, evidence indicates that non-enzymatic mechanisms also contribute. Specifically, the autoxidation of polyphenols has been identified as a secondary source of H_2O_2, suggesting that total peroxide activity depends not only on enzymatic capacity but also on the concentration and qualitative composition of phenolic constituents, which act as synergizing agents [122]. This dual mechanism may partially explain the broad-spectrum antimicrobial efficacy observed for Jarrah honey.

In addition to antibacterial effects, Jarrah honey demonstrates potent antifungal activity, including activity against dermatophytes [115]. As with antibacterial action, H_2O_2 alone appears insufficient to fully account for this antifungal potency, indicating synergistic interactions between peroxide generation and polyphenolic compounds [121]. These findings highlight the importance of considering the integrated chemical matrix of honey, rather than attributing bioactivity to a single compound or mechanism.

5.2.3. Impact on Intestinal Microbiota and Function

The intestinal microbiota has received increasing interest due to its crucial role in health. Dysbiosis, characterized by the proliferation of potentially pathogenic bacteria at the expense of probiotic species, can negatively affect intestinal immune function and barrier function, consequently leading to bowel changes. A 2020 study by Yuyuan Li and colleagues investigated the effects of Jarrah honey in a mouse model of constipation [123]. The mice were administered Jarrah honey (7.5 g/kg body weight) via gastric tube once daily for 5 days [123]. Researchers assessed fecal water content and intestinal transit rate, along with colonic concentrations of substance P (SP), vasoactive intestinal peptide (VIP), and serotonin (5-hydroxytryptamine; 5-HT) [123]. The study demonstrated that honey alleviates constipation by modulating the composition of the microbiota and promoting its eubiosis [123].

https://doi.org/10.3390/biotech15010005

5.3. *Agastache Honey*

Agastache honey is a monofloral honey derived from *Agastache rugosa*, a perennial herbaceous plant belonging to the Lamiaceae family. Also known as Korean mint or Korean hyssop, this species is native to Australia and various parts of Asia [124].

5.3.1. Chemical Signature and Authentication Markers

The distinctive chemical profile of Agastache honey is crucial for its authentication and therapeutic properties. The most abundant compounds identified include the phenylpropanoid derivative phenyllactic acid (148.3 mg/kg) and methyl syringate (46.1 mg/kg) [125]. Other phenolic compounds present in high concentrations are 4-hydroxybenzoic acid (11.8 mg/kg), *p*-coumaric acid, and gallic acid. The predominance and unique ratios of these compounds suggest their potential use as reliable chemical markers for authentication [125]. To validate the floral source, the honey's phenolic profile was compared directly with that of *A. rugosa* flower extracts. Compounds common to both the honey and the flowers included protocatechuic acid, 4-hydroxybenzoic acid, 2,4-dihydroxybenzoic acid, chlorogenic acid, vanillic acid, caffeic acid, syringic acid, phenyllactic acid, *p*-coumaric acid, ferulic acid, sinapic acid, methylsyringate, cinnamic acid, hesperetin, and kaempferol [125]. However, notable differences were observed in the quantification of several compounds, suggesting chemical transformation or selective transfer during the nectar-to-honey process. For instance, phenyllactic acid and methyl syringate were present at very high concentrations in honey but only at smaller concentrations in the flowers. Conversely, caffeic acid and chlorogenic acid were abundant in the flowers but only detected at low concentrations in the honey [125].

In a study by Anand and collaborators (2019) [125], several volatile compounds were reported as putative markers of Agastache honey, namely phenol, 2,4-bis(1,1-dimethylethyl) (12.77% of total volatiles), estragole (12.31%), nonanoic acid, ethyl ester (7.22%), 2-propenoic acid, 3-phenyl-, ethyl ester (6.32%), hexadecanoic acid, ethyl ester (5.68%), benzaldehyde, 4 methoxy (5.17%), β-caryophyllene (4.67%), nonanal (3.19%), and 2H-benzimidazol-2-one, 1,3-dihydro-5-methyl- (2.34%). In particular, estragole and phenol, 2,4-bis(1,1-dimethylethyl) were also correlated on the antifungal activity of the same honey, as discussed in the next paragraph. This association was supported by previous data indicating the antifungal efficacy of the two compounds [117]. The chemical structure of volatile and non-volatile markers of Agastache honey are reported in Figure 3.

5.3.2. Antimicrobial Mechanism and Comparative Potency

Agastache honey exhibits potent antimicrobial activity driven by a mechanism distinct from that of honeys like Manuka (Table 4). In a study by Anand and colleagues, the contribution of H_2O_2 to the overall antimicrobial effect was quantified [125]. While H_2O_2 was shown not to contribute to the activity of Manuka honey (which relies on MGO), it contributed significantly to other tested honeys, specifically 12% for Jarrah honey and 10% for Agastache honey [125]. Honeys derived from Leptospermum species (e.g., Manuka), as previously discussed, possess a primary mechanism based on the non-peroxide activity of MGO. The contribution of H_2O_2 to Agastache honey's action categorizes it predominantly as a peroxide Activity honey, where the combined effect of H_2O_2 and its rich phenolic profile drives its therapeutic potential [125].

https://doi.org/10.3390/biotech15010005

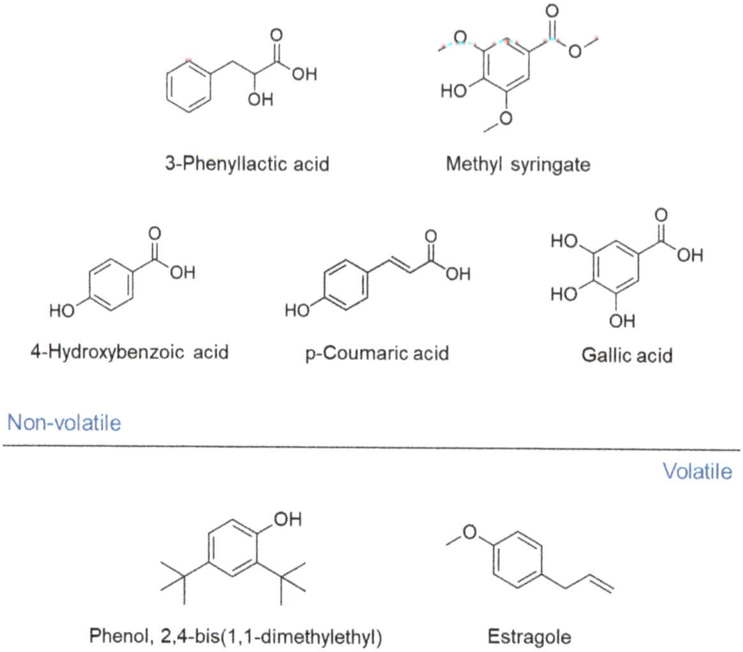

Figure 3. Chemical markers of Agastache honey.

Table 4. Exemplificative data showing the antibacterial activity of Agastache honey.

Targeted Microbe	Efficacy	Active Constituents	Ref.
Staphylococcus aureus (MSSA ATCC-25923)	MBC ≈ MIC: 6–25% *w/v*	Phenyllactic acid, methyl syringate; total phenolics	[125]
Staphylococcus aureus (MRSA ATCC BAA-1698)	MBC ≈ MIC: 6–25% *w/v*	Phenyllactic acid, methyl syringate; total phenolics	[125]
Escherichia coli (ATCC-11560 or clinical isolate)	MBC ≈ MIC: 6–25% *w/v*	H_2O_2; total phenolics	[125]
Pseudomonas aeruginosa (ATCC 21853 or clinical isolate)	MBC ≈ MIC: 6–25% *w/v*	H_2O_2; total phenolics	[125]
Trichophyton mentagrophytes	MFC: 40% *w/v*; zone diameter ~20 mm	H_2O_2; volatile terpenes, mainly estragole and phenol-2,4-bis (1,1-dimethylethyl)	[117]
Trichophyton rubrum	MFC: 40% *w/v*; zone diameter: 19.5 mm	H_2O_2; volatile terpenes, mainly estragole and phenol-2,4-bis (1,1-dimethylethyl)	[117]
Candida albicans (ATCC 10231 + clinical isolate)	MFC: 40% *w/v*	H_2O_2; volatile terpenes, mainly estragole and phenol-2,4-bis (1,1-dimethylethyl)	[117]

The ranges of MIC are related to different Agastache samples tested within the same study.

Agastache honey demonstrated particularly strong antifungal activity compared to both Manuka and Jarrah honeys when tested against the common dermatophytes *Trichophyton mentagrophytes* and *Trichophyton rubrum*, as well as the yeast *Candida albicans* [87]. Using agar diffusion and microdilution assays, Agastache honey was found to be effective at a concentration of 40% (*w/v*) against all tested fungal strains, yielding impressive zone diameters of 19.5–20 mm against the dermatophytes [87]. Crucially, the MIC and MFC values were identical at 40% against all tested strains [87]. This close equivalence

https://doi.org/10.3390/biotech15010005

between the MIC and MFC indicates a potent fungicidal action (i.e., the ability to kill the fungal cells, not just inhibit their growth). In contrast, Manuka honey only demonstrated fungistatic activity, requiring a higher 80% concentration against *T. mentagrophytes* and a 40% concentration against *T. rubrum* and *C. albicans* [87]. Jarrah honey, while showing some activity against *C. albicans*, exhibited no activity against the dermatophytes under the same assay conditions [87]. The superior antifungal efficacy of Agastache honey underscores its potential clinical utility for treating topical fungal infections [87].

5.4. Comparative Overview of Manuka, Jarrah, and Agastache Honeys

Manuka, Jarrah, and Agastache honeys exhibit overlapping yet distinct mechanistic profiles, bioactive compositions, and therapeutic potentials, which reflect their unique botanical and geographical origins. Manuka honey is characterized by a dominant non-peroxide mechanism, driven primarily by MGO, which confers robust antibacterial efficacy against Gram-positive and Gram-negative pathogens and underlines its clinical use in wound management. Conversely, Jarrah and Agastache honeys rely mainly on H_2O_2-mediated activity, with phenolic compounds acting synergistically to enhance antimicrobial and antifungal effects. While Jarrah shows broad-spectrum antibacterial activity and moderate antifungal potential, Agastache demonstrates a potent fungicidal activity, notably against dermatophytes and *C. albicans*, highlighting its potential for topical antifungal applications.

Despite these differences, all three honeys share bioactive convergence, including phenolic and flavonoid constituents that contribute to antioxidant, anti-inflammatory, and immunomodulatory effects. However, the relative contribution of individual components differs: MGO dominates in Manuka, specific polyphenols and unique volatiles in Jarrah, and phenyllactic acid, methyl syringate, and estragole in Agastache. These differences result in distinct chemical fingerprints that can serve both for authentication and mechanistic inference.

From a translational perspective, these mechanistic distinctions carry important implications for therapeutic applications. The MGO-driven activity of Manuka honey allows for consistent antibacterial performance across diverse wound microbiomes, while Jarrah and Agastache honeys may exhibit variable activity depending on H_2O_2 production and phenolic composition, which are influenced by floral source, harvest season, and storage. The pronounced antifungal potency of Agastache honey, despite modest peroxide contribution, underscores the role of specific phenolic and volatile constituents in augmenting efficacy, suggesting potential for development as a targeted antifungal agent.

6. Security-Related Issues

Globally, contamination by trace elements has declined due to more effective control of pollutant emissions and the implementation of stricter environmental regulations. Nevertheless, environmental pollution remains a persistent and significant public health concern. Urban and natural ecosystems continue to be threatened by industrial activities that are economically indispensable yet inherently generate substantial environmental contamination [126]. Laboratory studies have highlighted the adverse effects of metal and metalloid exposure on Western honeybees. At a historic mining site, bees were exposed to pollutants through dust, food, and water. More than 1000 bees collected from five apiaries located along an 11 km contamination gradient near a former gold mine in southern France were analyzed. The bees closest to the mine exhibited a 36% reduction in olfactory learning performance, and three-dimensional brain imaging revealed a 4% reduction in the volume of their olfactory centers, suggesting neurodevelopmental impairments linked to contaminant exposure [127,128].

https://doi.org/10.3390/biotech15010005

During foraging, bees can inadvertently gather environmental contaminants such as pesticides (herbicides, fungicides, and insecticides) along with nectar. Since honey reflects the chemical characteristics of the surrounding environment, it serves as a valuable indicator of overall ecosystem health. Pesticide residues detected in honey can be evaluated in relation to their potential impact on human health using maximum residue limits (MRLs). An MRL, expressed in milligrams of residue per kilogram of food, represents the highest permissible concentration of a pesticide considered legally acceptable by the Codex Alimentarius Commission for foods and feeds [129]. Human exposure to the herbicide glyphosate and its metabolite aminomethylphosphonic acid (AMPA) has been associated with metabolic disturbances and oxidative stress [130,131]. Epidemiological research has linked glyphosate exposure to several health disorders, including cancer [132], respiratory diseases [133], chronic kidney disease [134], neurological disorders, as well as metabolic dysfunction and oxidative stress. In the European Union, the MRL for glyphosate in honey is set at 0.05 ppm [135], whereas the New Zealand Ministry of Primary Industries allows a higher maximum value of 0.1 ppm, a discrepancy that has generated significant controversy. Recent reports highlight the detection of glyphosate above the legal limits in several honey samples from EU [136]; regarding Manuka honey from New Zealand, a recent independent has detected glyphosate residues above the National limits in some retail samples (https://nomoreglyphosate.nz/glyphosate-in-nz-honey-first-test-results/, accessed on 27 November 2025), although this has not been yet documented in the scientific literature. Nevertheless, these data indicate that the development of more rigorous and harmonized glyphosate regulations remains essential.

Heavy metal contamination in honey may originate from environmental pollution, certain beekeeping and agricultural practices, metal containers used during processing, and physicochemical characteristics of the soil and water at the honey's geographical origin. The most frequently detected heavy metals in honey include cadmium, lead, and arsenic [89]. Regulatory limits set maximum concentrations of 0.10 mg/kg for lead and 0.05 mg/kg for cadmium in honey [135], while the tolerable daily intake of arsenic is 3.0 μg/kg of body weight [137].

Hydroxymethylfurfural (HMF) is a cyclic aldehyde formed from the degradation of hexoses during prolonged storage, a process accelerated under acidic conditions, elevated temperatures, or via the Maillard reaction. Although not unique to honey, HMF is widespread in heat-processed foods containing sugars. The physicochemical properties of honey, including pH, acidity, fructose-glucose ratio, mineral content, and water activity, favor its formation. Since fresh honey contains little to no HMF, its concentration is widely used as an indicator of freshness and processing intensity. Storage conditions and the use of metal containers also influence HMF formation; therefore, minimizing heat exposure and storage duration is recommended [59]. To ensure product safety and quality, the Codex Alimentarius Commission has established a maximum HMF limit of 40 mg/kg for honey, with an increased limit of 80 mg/kg for honeys produced in tropical regions. Excessive intake of HMF has been linked to mutagenic, genotoxic, and organotoxic effects, with the compound demonstrating enzyme-inhibitory and DNA-damaging potential in multiple in vitro, in vivo, and preclinical studies [59]. After ingestion, HMF is readily absorbed from the gastrointestinal tract and subsequently metabolized into several derivatives excreted in urine. Interestingly, while HMF is associated with harmful effects, its metabolite 5-sulfoxymethylfurfural also exhibits beneficial biological activities, including antioxidant, antiallergic, anti-inflammatory, antihypoxic, anti-sickling, and antihyperuricemic properties. Following oral or intravenous administration, HMF is primarily converted into three metabolites [5-hydroxymethylfuroic acid (HMFA), 2,5-furandicarboxylic acid (FDCA), and 5-(hydroxymethyl)-2-furoyl glycine (HMFG)] and may also form a fourth product,

https://doi.org/10.3390/biotech15010005

5-sulfoxymethylfurfural (SMF) [138]. In bees, HMF is suspected to cause dysentery-like symptoms and gastrointestinal ulceration, ultimately contributing to mortality [139]. HMF content in commercial manuka and jarrah honeys from Oceania shows considerable variability, reflecting differences in storage and processing practices [110].

Pyrrolizidine alkaloids (PAs) and their N-oxides (PANOs) are secondary metabolic compounds produced by plants as chemical defenses against herbivores. They are classified into four groups based on the structure of their necine base, i.e., platynecine, retronecine, heliotridine, and otonecine. PAs are typically not found in free base form but occur as mono-, di-, or macrocyclic diesters formed by the esterification of a necine base with one or more aliphatic mono- or dicarboxylic acids, contributing to their structural diversity. The core pyrrolizidine skeleton consists of two fused saturated five-membered rings bridged by a nitrogen atom, and the presence of a 1,2-double bond often enhances toxicity. In their native form, PAs are protoxins and biologically inactive. To exert toxicity, they require metabolic activation, primarily in the liver, where CYP450 monooxygenases convert them into reactive electrophilic metabolites such as 6,7-dihydro-7-hydroxy-1-(hydroxymethyl)-5H-pyrrolizine. These metabolites readily bind to nucleophilic sites in DNA, proteins, and amino acids, forming pyrrole adducts that may persist in tissues and induce toxicity, particularly in the liver. PAs have been shown to possess hepatotoxic, pneumotoxic, genotoxic, carcinogenic, and developmental toxic effects. Detoxification can occur through conjugation with glutathione, producing more soluble and less toxic metabolites that are more easily excreted [135]. Recent studies in Australia have documented the presence of PAs in monofloral and multifloral honeys, including those produced by stingless bees. UHPLC-MS/MS analyses detected PAs such as lycopsamine, indicine, and intermedine, with plant origins traced to Parsonsia spp. and Ageratum conyzoides near apiaries [140]. These findings highlight that PA contamination remains a real safety concern even for premium medicinal honeys in Oceania. Exposure to PAs is considered one of the primary causes of hepatic sinusoidal obstruction syndrome (HSOS), formerly known as hepatic veno-occlusive disease, a rare but severe hepatic vascular disorder associated with high mortality [141,142].

7. Issues Related to Adulteration and Counterfeiting

Food fraud is a major concern for high-value medicinal honeys from Oceania, particularly Manuka and Jarrah. Because these honeys command premium prices and are prized for their therapeutic properties, they are especially attractive targets for adulteration and counterfeiting. In this context, fraud does not simply erode economic value, since it can undermine the bioactive properties that makes these honeys "medicinal." The most common form of adulteration involves the addition of inexpensive sugar syrups such as glucose, rice, corn, or maple syrup, to bulk up the product. Scientific studies have directly demonstrated this in Australian monofloral honeys. For example, using HPTLC, researchers successfully detected the addition of several syrup types (rice, corn, golden, maple, treacle) even at relatively low adulteration levels in both Manuka and Jarrah honeys [143]. Their chemometric analysis, combining sugar-profile and organic-extract data, allowed robust discrimination between authentic and adulterated samples [138].

Counterfeiting, in contrast, often involves misrepresentation of the botanical or geographic origin of honey, or the creation of synthetic honey analogues. Authenticating floral origin is critical for medicinal honeys because their value hinges on specific chemical markers. For instance, the recent article by Lawag and collaborators reporting the "nectar signature" for Jarrah honey that persist in the finished product provide a valuable tool to verify true Jarrah origin [116]. Despite these advances, longstanding analytical challenges remain, especially for manuka honey. The widely used AOAC 998.12 method (a stable

https://doi.org/10.3390/biotech15010005

carbon isotope test) can produce false positives in genuine Manuka. This is due to the conversion of DHA, naturally present in manuka nectar, to MGO over time or during storage, a reaction that shifts the $\delta^{13}C$ of the protein fraction and mimics the isotopic signature of C_4-sugar adulteration. Classic work has shown that insoluble components like pollen co-precipitate during protein isolation in this test, which further skews the $\delta^{13}C$ and may lead to misclassification. Adding to the challenge, pollen-based authentication (melissopalynology) can also be problematic. In Jarrah honey, for example, there is evidence that pollen analysis may misidentify closely related *Eucalyptus* species, making it unreliable as a sole indicator of floral origin. This underscores the need for combined chemical and botanical approaches.

More recently, researchers have moved toward multi-marker and chemometric techniques. The HPTLC-chemometric approach described earlier allowed discrimination of adulterated manuka and jarrah honeys even at low syrup addition percentages by using principal component analysis (PCA), cluster analysis, and artificial neural networks (ANN) [139,144,145]. These methods are particularly promising for defending medicinal honeys from fraud because they do not rely on just sugar composition but also track the organic extract fingerprint. Furthermore, elemental profiling has emerged as another powerful strategy. By analyzing the mineral content of honey and comparing it with known syrups, sPLS-DA (sparse Partial Least Squares Discriminant Analysis) models have achieved high accuracy in distinguishing pure honeys from adulterated ones [146,147]. This mineral-chemometric method could provide a stable and robust approach to detect syrup adulteration in a way that is less vulnerable to manipulation by counterfeiters.

The implications of adulteration and counterfeiting for medicinal honey integrity are profound. If a Manuka or Jarrah honey is diluted or mislabeled, its characteristic bioactive constituents may be reduced or masked, potentially diminishing therapeutic efficacy. To preserve the integrity of these medicinal honeys, research and industry must continue to develop and implement authentication frameworks. These should integrate isotopic methods, chemometric fingerprinting, nectar-marker profiling, and elemental analysis. Equally important is establishing regional reference libraries of authenticated honeys (with known botanical and geographic origin) to serve as benchmarks.

8. Conclusions

In recent years, honey has attracted considerable scientific interest for its potential beneficial properties, particularly anti-inflammatory, antimicrobial, antioxidant, and wound-healing activities, building on its traditional nutritional and medicinal use. Among honeys from Oceania, Manuka is by far the most extensively studied, with robust in vitro, preclinical, and some clinical evidence supporting its bioactivity and therapeutic potential. However, this preponderance of research has created a literature bias, leaving other medicinal honeys from the region, such as Jarrah and Agastache, comparatively under-characterized. Many endemic honeys remain unexplored, and their chemical composition, bioactive constituents, and potential clinical applications are largely unknown.

A critical limitation in the current literature is the high variability in reported bioactivities, reflecting differences in floral origin, geographical location, harvesting, and processing conditions. Standardization in honey characterization is frequently lacking, making comparisons between studies difficult and complicating efforts to translate in vitro findings into clinical practice. Moreover, the majority of antimicrobial and antifungal studies rely on planktonic bacterial models and high honey concentrations in vitro, which do not fully capture the complexity of human infections, particularly biofilm-associated or chronic wounds. The bioavailability of bioactive compounds at the tissue interface, as well as

https://doi.org/10.3390/biotech15010005

the impact of formulation (e.g., medical-grade honey, hydrogels, or dressings), are often overlooked, limiting our understanding of how laboratory results can be applied clinically.

Clinical evidence remains sparse, with most studies conducted in vitro or in preclinical models. This gap emphasizes the need for well-designed human trials to validate efficacy and safety, particularly for honeys beyond Manuka. Additionally, synergistic effects with conventional treatments, dosage optimization, and long-term safety profiles have not been systematically evaluated, further limiting clinical translation.

Future research should prioritize under-represented honeys, with comprehensive chemical and bioactivity profiling using standardized methods. Investigations should aim to uncover the molecular mechanisms responsible for observed bioactivities and assess their therapeutic potential in clinical settings. By addressing these gaps, researchers can expand the medicinal relevance of Oceania honeys, moving beyond the well-studied Manuka to fully exploit the region's rich apicultural biodiversity. Ultimately, such efforts could pave the way for novel natural therapeutics with well-defined efficacy, safety, and health-promoting properties.

Finally, as market demand increases, a greater tendency toward fraud has become prevalent, leading to the use of analytical tools for quality assessment. It is therefore essential to pay attention to honey's authenticity, seeking to develop techniques that can reliably determine its botanical origin to identify cases of adulteration and counterfeiting.

Author Contributions: Conceptualization, G.P.; methodology, G.P., M.L. and M.R.; validation, G.P.; formal analysis, G.P.; investigation, G.P., M.R. and M.L.; data curation, M.L.; writing—original draft preparation, M.R. and M.L.; writing—review and editing, G.P. and M.L.; visualization, G.P.; supervision, G.P.; project administration, G.P. All authors have read and agreed to the published version of the manuscript.

Funding: This research received no external funding.

Institutional Review Board Statement: Not applicable.

Informed Consent Statement: Not applicable.

Data Availability Statement: No new data were created or analyzed in this study.

Conflicts of Interest: The authors declare no conflicts of interest.

References

1. Legislative Decree No. 179 of 21 May 2004 on the Implementation of EC Directive 2001/110/EC Concerning the Production and Marketing of Honey; Italy. Available online: https://www.normattiva.it (accessed on 27 November 2025).
2. Lozada Lawag, I.; Green, K.J.; Khairul Islam, M.; Locher, C.; Hammer, K.A. Bioactivities and phenolic profiles of honeys derived from plants of the Goldfields, Esperance and Wheatbelt regions of Western Australia. *Chem. Biodivers.* **2023**, *20*, e202301678. [CrossRef] [PubMed]
3. Food and Agriculture Organization of the United Nations/WHO Codex Alimentarius. Available online: https://www.fao.org/fao-who-codexalimentarius/en/ (accessed on 27 November 2025).
4. COUNCIL DIRECTIVE 2001/110/EC of 20 December 2001 Relating to Honey. Available online: https://eur-lex.europa.eu/LexUriServ/LexUriServ.do?uri=OJ:L:2002:010:0047:0052:EN:PDF (accessed on 27 November 2025).
5. Ministero della Salute (Italy). Available online: https://www.salute.gov.it (accessed on 27 November 2025).
6. Escuredo, O.; Seijo, M.C. Honey: Chemical Composition, Stability and Authenticity. *Foods* **2019**, *8*, 577. [CrossRef]
7. Hunter, M.; Kellett, J.; Toohey, K.; Naumovski, N. Sensory and compositional properties affecting the likeability of commercially available Australian honeys. *Foods* **2021**, *10*, 1842. [CrossRef]
8. Eteraf-Oskouei, T.; Najafi, M. Traditional and modern uses of natural honey in human diseases: A review. *Iran. J. Basic Med. Sci.* **2013**, *16*, 731–742.
9. Schiassi, M.C.E.V.; de Souza, V.R.; Lago, A.M.T.; Carvalho, G.R.; Curi, P.N.; Guimarães, A.S.; Queiroz, F. Quality of honeys from different botanical origins. *J. Food Sci. Technol.* **2021**, *58*, 4167–4177. [CrossRef]

https://doi.org/10.3390/biotech15010005

10. Anklam, E. A review of the analytical methods to determine the geographical and botanical origin of honey. *Food Chem.* **1998**, *63*, 549–562. [CrossRef]
11. Truzzi, C.; Illuminati, S.; Annibaldi, A.; Finale, C.; Rossetti, M.; Scarponi, G. Physicochemical Properties of Honey from Marche, Central Italy: Classification of Unifloral and Multifloral Honeys by Multivariate Analysis. *Nat. Prod. Commun.* **2014**, *9*, 1595–1602. [CrossRef]
12. Gašić, U.M.; Milojković-Opsenica, D.M.; Tešić, Ž.L. Polyphenols as Possible Markers of Botanical Origin of Honey. *J. AOAC Int.* **2017**, *100*, 852–861. [CrossRef]
13. Wang, J.; Li, Q.X. Chemical composition, characterization, and differentiation of honey botanical and geographical origins. *Adv. Food Nutr. Res.* **2011**, *62*, 89–137. [CrossRef]
14. Vázquez, L.; Armada, D.; Celeiro, M.; Dagnac, T.; Llompart, M. Evaluating the presence and contents of phytochemicals in honey samples: Phenolic compounds as indicators to identify their botanical origin. *Foods* **2021**, *10*, 2616. [CrossRef] [PubMed]
15. Oddo, L.; Piana, L.; Sabatini, A.G. *Knowing Honey: A Guide to Sensory Analysis*; National Institute of Beekeeping: Bologna, Italy, 1997.
16. Surindro, A.; Takhellambam Chanu, M. Honey Bee Products: Honey and Royal Jelly and Their Nutritional and Medicinal Values to Humans. *Br. J. Biomed. Res.* **2021**, *5*, 1604–1619.
17. Persano Oddo, L.; Piazza, M.; Sabatini, A.; Accorti, M. Characterization of unifloral honeys. *Apidologie* **1995**, *26*, 453–465. [CrossRef]
18. Ranneh, Y.; Akim, A.M.; Hamid, H.A.; Khazaai, H.; Fadel, A.; Zakaria, Z.A.; Albujja, M.; Bakar, M.F.A. Honey and its nutritional and anti-inflammatory value. *BMC Complement. Med. Ther.* **2021**, *21*, 30. [CrossRef]
19. Nguyen, H.T.L.; Panyoyai, N.; Kasapis, S.; Pang, E.; Mantri, N. Honey and its role in relieving multiple facets of atherosclerosis. *Nutrients* **2019**, *11*, 167. [CrossRef]
20. Miguel, M.G.; Antunes, M.D.; Faleiro, M.L. Honey as a complementary medicine. *Integr. Med. Insights* **2017**, *12*, 1178633717702869. [CrossRef] [PubMed]
21. Pasupuleti, V.R.; Sammugam, L.; Ramesh, N.; Gan, S.H. Honey, Propolis, and Royal Jelly: A Comprehensive Review of Their Biological Actions and Health Benefits. *Oxid. Med. Cell. Longev.* **2017**, *2017*, 1259510. [CrossRef] [PubMed] [PubMed Central]
22. Da Silva, P.M.; Gauche, C.; Gonzaga, L.V.; Costa, A.C.; Fett, R. Honey: Chemical composition, stability and authenticity. *Food Chem.* **2016**, *196*, 309–323. [CrossRef] [PubMed]
23. Nguyen, H.T.L.; Panyoyai, N.; Paramita, V.D.; Mantri, N.; Kasapis, S. Physicochemical and viscoelastic properties of honey from medicinal plants. *Food Chem.* **2018**, *241*, 143–149. [CrossRef] [PubMed]
24. Atkinson, F.S.; Foster-Powell, K.; Brand-Miller, J.C. International tables of glycemic index and glycemic load values: 2008. *Diabetes Care* **2008**, *31*, 2281–2283. [CrossRef]
25. Bogdanov, S.; Jurendic, T.; Sieber, R.; Gallmann, P. Honey for nutrition and health: A review. *J. Am. Coll. Nutr.* **2008**, *27*, 677–689. [CrossRef]
26. Ball, D.W. The chemical composition of honey. *J. Chem. Educ.* **2007**, *84*, 1643–1646. [CrossRef]
27. Da Costa Leite, J.M.; Trugo, L.C.; Costa, L.S.M.; Quinteiro, L.M.C.; Barth, O.M.; Dutra, V.M.L.; De Maria, C.A.B. Determination of oligosaccharides in Brazilian honeys of different botanical origin. *Food Chem.* **2000**, *70*, 93–98. [CrossRef]
28. De La Fuente, E.; Ruiz-Matute, A.I.; Valencia-Barrera, R.M.; Sanz, J.; Martínez Castro, I. Carbohydrate composition of Spanish unifloral honeys. *Food Chem.* **2011**, *129*, 1483–1489. [CrossRef]
29. Lazarević, K.B.; Jovetić, M.S.; Tešić, Ž.L. Physicochemical parameters as a tool for the assessment of origin of honey. *J. AOAC Int.* **2017**, *100*, 840–851. [CrossRef] [PubMed]
30. Tedesco, R.; Scalabrin, E.; Malagnini, V.; Strojnik, L.; Ogrinc, N.; Capodaglio, G. Characterization of botanical origin of Italian honey by carbohydrate composition and volatile organic compounds (VOCs). *Foods* **2022**, *11*, 2441. [CrossRef]
31. Zammit Young, G.-W.; Blundell, R. A review on the phytochemical composition and health applications of honey. *Heliyon* **2023**, *9*, e12507. [CrossRef] [PubMed]
32. Kumar, S.; Pandey, A.K. Chemistry and biological activities of flavonoids: An overview. *Sci. World J.* **2013**, *2013*, 162750. [CrossRef]
33. Cushnie, T.P.T.; Lamb, A.J. Antimicrobial activity of flavonoids. *Int. J. Antimicrob. Agents* **2005**, *26*, 343–356. [CrossRef]
34. Wilczyńska, A.; Żak, N. Polyphenols as the main compounds influencing the antioxidant effect of honey—A review. *Int. J. Mol. Sci.* **2024**, *25*, 10606. [CrossRef]
35. Harborne, J.B.; Williams, C.A. Advances in flavonoid research since 1992. *Phytochemistry* **2000**, *55*, 481–504. [CrossRef]
36. Beretta, G.; Granata, P.; Ferrero, M.; Orioli, M.; Facino, R.M. Standardization of antioxidant properties of honey by a combination of spectrophotometric/fluorimetric assays and chemometrics. *Anal. Chim. Acta* **2005**, *523*, 185–191. [CrossRef]
37. Vijaya, K.; Ananthan, S. Therapeutic efficacy of medicinal plants against experimentally induced shigellosis in guinea pigs. *Indian. J. Pharm. Sci.* **1996**, *58*, 191–193.
38. Lu, Y.; Yang, J.; Wang, X.; Ma, Z.; Li, S.; Liu, Z.; Fan, X. Research progress in use of traditional Chinese medicine for treatment of spinal cord injury. *Biomed. Pharmacother.* **2020**, *127*, 110136. [CrossRef]

 https://doi.org/10.3390/biotech15010005

39. Ikigai, H.; Nakae, T.; Hara, Y.; Shimamura, T. Bactericidal catechins damage the lipid bilayer. *Biochim. Biophys. Acta* **1993**, *1147*, 132–136. [CrossRef] [PubMed]
40. Cushnie, T.P.; Lamb, A.J. Detection of galangin-induced cytoplasmic membrane damage in Staphylococcus aureus by measuring potassium loss. *J. Ethnopharmacol.* **2005**, *101*, 243–248. [CrossRef]
41. Elbatreek, M.H.; Mahdi, I.; Ouchari, W.; Mahmoud, M.F.; Sobeh, M. Current advances on the therapeutic potential of pinocembrin: An updated review. *Biomed. Pharmacother.* **2023**, *157*, 114032. [CrossRef] [PubMed]
42. Nguyen, T.L.A.; Bhattacharya, D. Antimicrobial Activity of Quercetin: An Approach to Its Mechanistic Principle. *Molecules* **2022**, *27*, 2494. [CrossRef] [PubMed]
43. Yue, B.; Ren, J.; Yu, Z.; Luo, X.; Ren, Y.; Zhang, J.; Mani, S.; Wang, Z.; Dou, W. Pinocembrin alleviates ulcerative colitis in mice via regulating gut microbiota, suppressing TLR4/MD2/NF-κB pathway and promoting intestinal barrier. *Biosci. Rep.* **2020**, *40*, BSR20200986. [CrossRef]
44. Gong, L.J.; Wang, X.Y.; Gu, W.Y.; Wu, X. Pinocembrin ameliorates intermittent hypoxia-induced neuroinflammation through BNIP3-dependent mitophagy in a murine model of sleep apnea. *J. Neuroinflamm.* **2020**, *17*, 337. [CrossRef]
45. Pei, B.; Sun, J. Pinocembrin alleviates cognition deficits by inhibiting inflammation in diabetic mice. *J. Neuroimmunol.* **2018**, *314*, 42–49. [CrossRef]
46. Liu, R.; Li, J.-Z.; Song, J.-K.; Sun, J.-L.; Li, Y.-J.; Zhou, S.-B.; Zhang, T.-T.; Du, G.-H. Pinocembrin protects human brain microvascular endothelial cells against fibrillar amyloid-β1–40 injury by suppressing the MAPK/NF-κB inflammatory pathways. *Biomed. Res. Int.* **2014**, *2014*, 470393. [PubMed]
47. Wang, W.; Zheng, L.; Xu, L.; Tu, J.; Gu, X. Pinocembrin mitigates depressive-like behaviors induced by chronic unpredictable mild stress through ameliorating neuroinflammation and apoptosis. *Mol. Med.* **2020**, *26*, 53. [CrossRef]
48. Cheung, Y.; Meenu, M.; Yu, X.; Xu, B. Phenolic acids and flavonoids profiles of commercial honey from different floral sources and geographic sources. *Int. J. Food Prop.* **2019**, *22*, 290–308. [CrossRef]
49. Ciulu, M.; Spano, N.; Pilo, M.; Sanna, G. Recent advances in the analysis of phenolic compounds in unifloral honeys. *Molecules* **2016**, *21*, 451. [CrossRef]
50. Arceusz, A.; Wesolowski, M.; Konieczynski, P. Methods for extraction and determination of phenolic acids in medicinal plants: A review. *Nat. Prod. Commun.* **2013**, *8*, 1821–1829. [CrossRef]
51. Trautvetter, S.; Koelling-Speer, I.; Speer, K. Confirmation of phenolic acids and flavonoids in honeys by UPLC-MS. *Apidologie* **2009**, *40*, 140–150. [CrossRef]
52. Majewska, E.; Drużyńska, B.; Derewiaka, D.; Ciecierska, M.; Pakosz, P. Comparison of antioxidant properties and color of selected Polish honeys and manuka honey. *Foods* **2024**, *13*, 2666. [CrossRef]
53. Seraglio, S.K.T.; Bergamo, G.; Brugnerotto, P.; Gonzaga, L.V.; Fett, R.; Costa, A.C.O. Aliphatic organic acids as promising authenticity markers of bracatinga honeydew honey. *Food Chem.* **2021**, *343*, 128449. [CrossRef]
54. Brugnerotto, P.; Della Betta, F.; Gonzaga, L.V.; Fett, R.; Oliveira Costa, A.C. A capillary electrophoresis method to determine aliphatic organic acids in bracatinga honeydew honey and floral honey. *J. Food Compos. Anal.* **2019**, *82*, 103243. [CrossRef]
55. Liao, Y.; Zhang, Z.; Zhao, Y.; Zhang, S.; Zha, K.; Ouyang, L.; Hu, W.; Zhou, W.; Sun, Y.; Liu, G. Glucose oxidase: An emerging multidimensional treatment option for diabetic wound healing. *Bioact. Mater.* **2024**, *44*, 131–151. [CrossRef] [PubMed]
56. Bankar, S.B.; Bule, M.V.; Singhal, R.S.; Ananthanarayan, L. Glucose oxidase—An overview. *Biotechnol. Adv.* **2009**, *27*, 489–501. [CrossRef]
57. Ligor, M.; Kowalkowski, T.; Buszewski, B. Comparative study of the potentially toxic elements and essential microelements in honey depending on the geographic origin. *Molecules* **2022**, *27*, 5474. [CrossRef] [PubMed]
58. Solayman, M.; Islam, M.A.; Paul, S.; Ali, Y.; Khalil, M.I.; Alam, N.; Gan, S.H. Physicochemical properties, minerals, trace elements, and heavy metals in honey of different origins: A comprehensive review. *Compr. Rev. Food Sci. Food Saf.* **2016**, *15*, 219–233. [CrossRef]
59. Shapla, U.M.; Solayman, M.; Alam, N.; Khalil, M.I.; Gan, S.H. 5-Hydroxymethylfurfural (HMF) levels in honey and other food products: Effects on bees and human health. *Chem. Cent. J.* **2018**, *12*, 35.
60. Weis, W.A.; Ripari, N.; Conte, F.L.; Honorio, M.d.S.; Sartori, A.A.; Matucci, R.H.; Sforcin, J.M. An overview about apitherapy and its clinical applications. *Phytomed. Plus* **2022**, *2*, 100239.
61. Erejuwa, O.O.; Sulaiman, S.A.; Wahab, M.S.; Sirajudeen, K.N.; Salleh, M.S.; Gurtu, S. Antioxidant protection of Malaysian tualang honey in pancreas of normal and streptozotocin-induced diabetic rats. *Ann. Endocrinol.* **2010**, *71*, 291–296. [CrossRef] [PubMed]
62. Erejuwa, O.O.; Sulaiman, S.A.; Ab Wahab, M.S. Honey: A novel antioxidant. *Molecules* **2012**, *17*, 4400–4423. [CrossRef] [PubMed]
63. Rakha, M.K.; Nabil, Z.I.; Hussein, A.A. Cardioactive and vasoactive effects of natural wild honey against cardiac malperformance induced by hyperadrenergic activity. *J. Med. Food* **2008**, *11*, 91–98. [CrossRef]
64. Hassan, M.I.; Mabrouk, G.M.; Shehata, H.H.; Aboelhussein, M.M. Antineoplastic effects of bee honey and Nigella sativa on hepatocellular carcinoma cells. *Integr. Cancer Ther.* **2012**, *11*, 354–363. [CrossRef]
65. Almasaudi, S. The antibacterial activities of honey. *Saudi J. Biol. Sci.* **2021**, *28*, 2188–2196. [CrossRef]

 https://doi.org/10.3390/biotech15010005

66. Hossain, M.L.; Lim, L.Y.; Hammer, K.; Hettiarachchi, D.; Locher, C. A review of commonly used methodologies for assessing the antibacterial activity of honey and honey products. *Antibiotics* **2022**, *11*, 975. [CrossRef]
67. Olaitan, P.B.; Adeleke, O.E.; Ola, I.O. Honey: A reservoir for microorganisms and an inhibitory agent for microbes. *Afr. Health Sci.* **2007**, *7*, 159–165.
68. Ayefoumi Adinortey, C.; Wilson, M.; Kojo Kwofie, S. Honey as a natural product worthy of re-consideration in treating MRSA wound infections. In *The Global Antimicrobial Resistance Epidemic—Innovative Approaches and Cutting-Edge Solutions*; Tellez-Isaias, G., Ed.; IntechOpen: London, UK, 2022. [CrossRef]
69. Shokri, H.; Sharifzadeh, A. Fungicidal efficacy of various honeys against fluconazole-resistant Candida species isolated from HIV+ patients with candidiasis. *J. Mycol. Méd.* **2017**, *27*, 159–165. [CrossRef]
70. Rocha, M.P.; Amorim, J.M.; Lima, W.G.; Brito, J.C.M.; da Cruz Nizer, W.S. Effect of honey and propolis, compared to acyclovir, against herpes simplex virus (HSV)-induced lesions: A systematic review and meta-analysis. *J. Ethnopharmacol.* **2022**, *287*, 114939. [CrossRef]
71. Silva, B.; Biluca, F.C.; Gonzaga, L.V.; Fett, R.; Dalmarco, E.M.; Caon, T.; Costa, A.C.O. In vitro anti-inflammatory properties of honey flavonoids: A review. *Food Res. Int.* **2021**, *141*, 110086. [CrossRef]
72. Candiracci, M.; Piatti, E.; Dominguez-Barragán, M.; García-Antrás, D.; Morgado, B.; Ruano, D.; Gutiérrez, J.F.; Parrado, J.; Castaño, A. Anti-inflammatory activity of a honey flavonoid extract on lipopolysaccharide-activated N13 microglial cells. *J. Agric. Food Chem.* **2012**, *60*, 12304–12311. [CrossRef] [PubMed]
73. Al-Waili, N.S.; Boni, N.S. Natural honey lowers plasma prostaglandin concentrations in normal individuals. *J. Med. Food* **2003**, *6*, 129–133. [CrossRef] [PubMed]
74. Subrahmanyam, M. A prospective randomized clinical and histological study of superficial burn wound healing with honey and silver sulfadiazine. *Burns* **1998**, *24*, 157–161. [CrossRef] [PubMed]
75. Yupanqui Mieles, J.; Vyas, C.; Aslan, E.; Humphreys, G.; Diver, C.; Bartolo, P. Honey: An advanced antimicrobial and wound healing biomaterial for tissue engineering applications. *Pharmaceutics* **2022**, *14*, 1663. [CrossRef]
76. Działo, M.; Mierziak, J.; Korzun, U.; Preisner, M.; Szopa, J.; Kulma, A. The potential of plant phenolics in prevention and therapy of skin disorders. *Int. J. Mol. Sci.* **2016**, *17*, 160. [CrossRef]
77. Safitri, E.; Purnobasuki, H.; Al-Anshori, A.A.; Lestari, T.D.; Utama, S.; Yimer, N. Honey-derived phytochemicals: Implications for stem cell activation and health benefits. *J. Funct. Foods* **2024**, *114*, 106058. [CrossRef]
78. Ahmed, S.; Othman, N.H. Honey as a potential natural anticancer agent: A review of its mechanisms. *Evid.-Based Complement. Altern. Med.* **2013**, *2013*, 829070. [CrossRef]
79. Majtán, J.; Kováčová, E.; Bíliková, K.; Šimúth, J. The immunostimulatory effect of the recombinant apalbumin 1—Major honeybee royal jelly protein—On TNF-α release. *Int. Immunopharmacol.* **2006**, *6*, 269–278. [CrossRef]
80. Chan-Zapata, I.; Segura-Campos, M.R. Honey and its protein components: Effects in the cancer immunology. *J. Food Biochem.* **2021**, *45*, e13613. [CrossRef]
81. MacEwan, D.J. TNF ligands and receptors—A matter of life and death. *Br. J. Pharmacol.* **2002**, *135*, 855–875. [CrossRef] [PubMed]
82. Mavrič, E.; Wittmann, S.; Barth, G.; Henle, T. Identification and quantification of methylglyoxal as the dominant antibacterial constituent of manuka (*Leptospermum scoparium*) honeys from New Zealand. *Mol. Nutr. Food Res.* **2008**, *52*, 483–489. [CrossRef]
83. Roberts, A.E.L.; Maddocks, S.E.; Cooper, R.A. Manuka honey is bactericidal against Pseudomonas aeruginosa and results in differential expression of oprF and algD. *Microbiology* **2012**, *158*, 3005–3013. [CrossRef] [PubMed]
84. Nolan, V.C.; Harrison, J.; Wright, J.E.E.; Cox, J.A.G. Clinical significance of manuka and medical-grade honey for antibiotic-resistant infections: A systematic review. *Antibiotics* **2020**, *9*, 766. [CrossRef]
85. Juliano, C.; Magrini, G.A. Methylglyoxal, the major antibacterial factor in manuka honey: An alternative to preserve natural cosmetics? *Cosmetics* **2019**, *6*, 1. [CrossRef]
86. Henriques, A.F.; Jenkins, R.E.; Burton, N.F.; Cooper, R.A. The intracellular effects of manuka honey on Staphylococcus aureus. *Eur. J. Clin. Microbiol. Infect. Dis.* **2010**, *29*, 45–50. [CrossRef]
87. Lawag, I.L.; Lim, L.-Y.; Joshi, R.; Hammer, K.A.; Locher, C. A comprehensive survey of phenolic constituents reported in monofloral honeys around the globe. *Foods* **2022**, *11*, 1152. [CrossRef]
88. El-Senduny, F.F.; Hegazi, N.M.; Abd Elghani, G.E.; Farag, M.A. Manuka honey, a unique monofloral honey: A comprehensive review of its bioactives, metabolism, action mechanisms, and therapeutic merits. *Food Biosci.* **2021**, *42*, 101038. [CrossRef]
89. Mello Dos Santos, M.; Khan, N.; Lim, L.Y.; Locher, C. Antioxidant activity, physicochemical and sensory properties of stingless bee honey from Australia. *Foods* **2024**, *13*, 1657. [CrossRef]
90. Smallfield, B.M.; Joyce, N.I.; Van Klink, J.W. Developmental and compositional changes in Leptospermum scoparium nectar and their relevance to mānuka honey bioactives and markers. *N. Z. J. Bot.* **2018**, *56*, 183–197. [CrossRef]
91. Dahiya, D.; Mackin, C.; Nigam, P.S. Studies on bioactivities of manuka and regional varieties of honey for their potential use as natural antibiotic agents for infection control related to wound healing and in pharmaceutical formulations. *AIMS Microbiol.* **2024**, *10*, 288–310. [CrossRef]

92. Camplin, A.L.; Maddocks, S.E. Manuka honey treatment of biofilms of Pseudomonas aeruginosa results in the emergence of isolates with increased honey resistance. *Ann. Clin. Microbiol. Antimicrob.* **2014**, *13*, 19. [CrossRef]
93. Lin, S.M.; Molan, P.C.; Cursons, R.T. The controlled in vitro susceptibility of gastrointestinal pathogens to the antibacterial effect of manuka honey. *Eur. J. Clin. Microbiol. Infect. Dis.* **2011**, *30*, 569–574. [CrossRef]
94. Girma, A.; Seo, W.; She, R.C. Antibacterial activity of varying UMF-graded manuka honeys. *PLoS ONE* **2019**, *14*, e0224495. [CrossRef]
95. Rabie, E.; Serem, J.C.; Oberholzer, H.M.; Gaspar, A.R.M.; Bester, M.J. How methylglyoxal kills bacteria: An ultrastructural study. *Ultrastruct. Pathol.* **2016**, *40*, 107–111. [CrossRef]
96. Feng, Y. Antibacterial properties of manuka honey and the role of methylglyoxal. *J. Stud. Res.* **2023**, 1–12. [CrossRef]
97. Sęk, A.; Porębska, A.; Szczęsna, T. Quality of commercially available manuka honey expressed by pollen composition, diastase activity, and hydroxymethylfurfural content. *Foods* **2023**, *12*, 2930. [CrossRef]
98. Pucci, M.; Aria, F.; Premoli, M.; Maccarinelli, G.; Mastinu, A.; Bonini, S.; Memo, M.; Uberti, D.; Abate, G. Methylglyoxal affects cognitive behavior and modulates RAGE and presenilin-1 expression in hippocampus of aged mice. *Food Chem. Toxicol.* **2021**, *158*, 112608.
99. Terio, V.; Bozzo, G.; Ceci, E.; Savarino, A.E.; Barrasso, R.; Di Pinto, A.; Mottola, A.; Marchetti, P.; Tantillo, G.; Bonerba, E. Methylglyoxal (MGO) in Italian honey. *Appl. Sci.* **2021**, *11*, 831. [CrossRef]
100. Li, X.; Prebble, J.G.; de Lange, P.J.; Raine, J.I.; Newstrom-Lloyd, L. Discrimination of pollen of New Zealand mānuka (*Leptospermum scoparium* agg.) and kānuka (*Kunzea* spp.) (Myrtaceae). *PLoS ONE* **2022**, *17*, e0269361. [CrossRef] [PubMed]
101. Hegazi, N.M.; Elghani, G.E.A.; Farag, M.A. The super-food manuka honey, a comprehensive review of its analysis and authenticity approaches. *J. Food Sci. Technol.* **2022**, *59*, 2527–2534. [CrossRef]
102. Hermanns, R.; Mateescu, C.; Thrasyvoulou, A.; Tananaki, C.; Wagener, F.A.D.T.G.; Cremers, N.A.J. Defining the standards for medical grade honey. *J. Apic. Res.* **2019**, *59*, 125–135. [CrossRef]
103. Álvarez-Suárez, J.; Gasparrini, M.; Forbes-Hernández, T.; Mazzoni, L.; Giampieri, F. The composition and biological activity of honey: A focus on manuka honey. *Foods* **2014**, *3*, 420–432. [CrossRef]
104. Al-Kafaween, M.A.; Al-Groom, R.M.; Mohd Hilmi, A.B. Comparison of the antimicrobial and antivirulence activities of Sidr and Tualang honeys with manuka honey against Staphylococcus aureus. *Iran. J. Microbiol.* **2023**, *15*, 89–101. [CrossRef]
105. Jenkins, R.E.; Cooper, R. Synergy between oxacillin and manuka honey sensitizes methicillin-resistant Staphylococcus aureus to oxacillin. *J. Antimicrob. Chemother.* **2012**, *67*, 1405–1407. [CrossRef]
106. Müller, P.; Alber, D.G.; Turnbull, L.; Schlothauer, R.C.; Carter, D.A.; Whitchurch, C.B.; Harry, E.J. Synergism between Medihoney and rifampicin against methicillin-resistant Staphylococcus aureus (MRSA). *PLoS ONE* **2013**, *8*, e57679. [CrossRef] [PubMed]
107. Brites, A.; Ferreira, M.; Bom, S.; Grenho, L.; Claudio, R.; Gomes, P.S.; Fernandes, M.H.; Marto, J.; Santos, C. Fabrication of antibacterial and biocompatible 3D printed manuka-gelatin based patch for wound healing applications. *Int. J. Pharm.* **2023**, *632*, 122541. [CrossRef] [PubMed]
108. Tomić, S.L.; Vuković, J.S.; Radić, M.M.B.; Filipović, V.V.; Živanović, D.P.; Nikolić, M.M.; Nikodinovic-Runic, J. Manuka honey/2-hydroxyethyl methacrylate/gelatin hybrid hydrogel scaffolds for potential tissue regeneration. *Polymers* **2023**, *15*, 589. [CrossRef]
109. Hossain, M.L.; Lim, L.Y.; Hammer, K.; Hettiarachchi, D.; Locher, C. Determination of antioxidant and antibacterial activities of honey-loaded topical formulations: A focus on Western Australian honeys. *Appl. Sci.* **2023**, *13*, 7440. [CrossRef]
110. Wang, S.; Qiu, Y.; Zhu, F. An updated review of functional ingredients of manuka honey and their value-added innovations. *Food Chem.* **2024**, *440*, 138060. [CrossRef]
111. Nakamura, A.; Kawaharada, R. Advanced glycation end products and oxidative stress in a hyperglycaemic environment. In *Fundamentals of Glycosylation*; Raghav, A., Ahmad, J., Eds.; IntechOpen: London, UK, 2022. [CrossRef]
112. Sahi, A.K.; Verma, P.; Varshney, N.; Gundu, S.; Mahto, S.K. Revisiting methodologies for in vitro preparations of advanced glycation end products. *Appl. Biochem. Biotechnol.* **2022**, *194*, 2831–2855. [CrossRef] [PubMed]
113. Vašková, J.; Kováčová, G.; Pudelský, J.; Palenčár, D.; Mičková, H. Methylglyoxal formation—Metabolic routes and consequences. *Antioxidants* **2025**, *14*, 212. [CrossRef]
114. Campo, I.; Morbini, P.; Pozzi, E.; Luisetti, M. RAGE: Una nuova via pro-infiammatoria per le polmoniti interstiziali? [AGE receptor: A novel pro-inflammatory pathway for chronic respiratory disorders?]. *Recenti Prog. Med.* **2008**, *99*, 271–278.
115. Islam, M.K.; Barbour, E.; Locher, C. Authentication of jarrah (*Eucalyptus marginata*) honey through its nectar signature and assessment of its typical physicochemical characteristics. *PeerJ Anal. Chem.* **2024**, *6*, e33. [CrossRef]
116. Lawag, I.L.; Islam, M.K.; Sostaric, T.; Lim, L.Y.; Hammer, K.; Locher, C. Antioxidant activity and phenolic compound identification and quantification in Western Australian honeys. *Antioxidants* **2023**, *12*, 189. [CrossRef]
117. Anand, S.; Deighton, M.; Livanos, G.; Pang, E.C.K.; Mantri, N. Agastache honey has superior antifungal activity in comparison with important commercial honeys. *Sci. Rep.* **2019**, *9*, 18197. [CrossRef]

118. Jones, Z.J.M.; Huang, Y.; Green, K.J.; Hammer, K.A. Changes in antibacterial activity, colour, and hydrogen peroxide content of Western Australian jarrah and marri honeys after storage at different temperatures over time. *J. Appl. Microbiol.* **2023**, *134*, lxad164. [CrossRef] [PubMed]
119. Green, K.J.; Dods, K.; Hammer, K.A. Development and validation of a new microplate assay that utilises optical density to quantify the antibacterial activity of honeys including jarrah, marri and manuka. *PLoS ONE* **2020**, *15*, e0243246. [CrossRef]
120. Sindi, A.; Chawn, M.V.B.; Hernandez, M.E.; Green, K.; Islam, M.K.; Locher, C.; Hammer, K. Anti-biofilm effects and characterisation of the hydrogen peroxide activity of a range of Western Australian honeys compared to manuka and multifloral honeys. *Sci. Rep.* **2019**, *9*, 17666. [CrossRef]
121. Guttentag, A.; Krishnakumar, K.; Cokcetin, N.; Hainsworth, S.; Harry, E.; Carter, D. Inhibition of dermatophyte fungi by Australian jarrah honey. *Pathogens* **2021**, *10*, 194. [CrossRef]
122. Brudzynski, K. Unexpected value of honey color for prediction of a non-enzymatic H2O2 production and honey antibacterial activity: A perspective. *Metabolites* **2023**, *13*, 526. [CrossRef]
123. Li, Y.; Long, S.; Liu, Q.; Ma, H.; Li, J.; Xiaoqing, W.; Yuan, J.; Li, M.; Hou, B. Gut microbiota is involved in the alleviation of loperamide-induced constipation by honey supplementation in mice. *Food Sci. Nutr.* **2020**, *8*, 4388–4398. [CrossRef]
124. Gong, H.; Li, S.; He, L.; Kasimu, R. Microscopic identification and in vitro activity of *Agastache rugosa* (Fisch. et Mey) from Xinjiang, China. *BMC Complement. Altern. Med.* **2017**, *17*, 95. [CrossRef]
125. Anand, S.; Deighton, M.; Livanos, G.; Morrison, P.D.; Pang, E.C.K.; Mantri, N. Antimicrobial activity of Agastache honey and characterization of its bioactive compounds in comparison with important commercial honeys. *Front. Microbiol.* **2019**, *10*, 263. [CrossRef]
126. Taylor, M.P.; Gillings, M.M.; Fry, K.L.; Barlow, C.F.; Gunkel-Grillion, P.; Gueyte, R.; Camoin, M. Tracing nickel smelter emissions using European honeybees. *Environ. Pollut.* **2023**, *335*, 122257. [CrossRef] [PubMed]
127. Monchanin, C.; Drujont, E.; Le Roux, G.; Lösel, P.D.; Barron, A.B.; Devaud, J.-M.; Elger, A.; Lihoreau, M. Environmental exposure to metallic pollution impairs honeybee brain development and cognition. *J. Hazard. Mater.* **2024**, *465*, 133218. [CrossRef] [PubMed]
128. Roncada, P.; Isani, G.; Peloso, M.; Dalmonte, T.; Bonan, S.; Caprai, E. Pyrrolizidine alkaloids from monofloral and multifloral Italian honey. *Int. J. Environ. Res. Public Health* **2023**, *20*, 5410. [CrossRef]
129. Boakye, R.G.; Stanley, D.A.; White, B. Honey contamination from plant protection products approved for cocoa (*Theobroma cacao*) cultivation: A systematic review of existing research and methods. *PLoS ONE* **2023**, *18*, e0280175. [CrossRef] [PubMed]
130. Cosemans, C.; Van Larebeke, N.; Janssen, B.G.; Martens, D.S.; Baeyens, W.; Bruckers, L.; Den Hond, E.; Coertjens, D.; Nelen, V.; Schoeters, G.; et al. Glyphosate and AMPA exposure in relation to markers of biological aging in an adult population-based study. *Int. J. Hyg. Environ. Health* **2022**, *240*, 113895. [CrossRef]
131. Leon, M.E.; Schinasi, L.H.; LeBailly, P.; Freeman, L.E.B.; Nordby, K.-C.; Ferro, G.; Monnereau, A.; Brouwer, M.; Tual, S.; Baldi, I.; et al. Pesticide use and risk of non-Hodgkin lymphoid malignancies in agricultural cohorts from France, Norway and the USA: A pooled analysis from the AGRICOH consortium. *Int. J. Epidemiol.* **2019**, *48*, 1519–1535. [CrossRef]
132. Hoppin, J.A.; Umbach, D.M.; London, S.J.; Henneberger, P.K.; Kullman, G.J.; Alavanja, M.C.R.; Sandler, D.P. Pesticides and atopic and nonatopic asthma among farm women in the Agricultural Health Study. *Am. J. Respir. Crit. Care Med.* **2008**, *177*, 11–18. [CrossRef]
133. Gunatilake, S.; Seneff, S.; Orlando, L. Glyphosate's synergistic toxicity in combination with other factors as a cause of chronic kidney disease of unknown origin. *Int. J. Environ. Res. Public Health* **2019**, *16*, 2734. [CrossRef] [PubMed]
134. von Ehrenstein, O.S.; Ling, C.; Cui, X.; Cockburn, M.; Park, A.S.; Yu, F.; Wu, J.; Ritz, B. Prenatal and infant exposure to ambient pesticides and autism spectrum disorder in children: Population-based case-control study. *BMJ* **2019**, *364*, l962. [CrossRef] [PubMed]
135. Rampazzo, G.; Gazzotti, T.; Zironi, E.; Pagliuca, G. Glyphosate and glufosinate residues in honey and other hive products. *Foods* **2023**, *12*, 1155. [CrossRef]
136. Kitts, D.D.; Chen, X.-M.; Jing, H. Demonstration of antioxidant and anti-inflammatory bioactivities from sugar–amino acid Maillard reaction products. *J. Agric. Food Chem.* **2012**, *60*, 6718–6727. [CrossRef]
137. Islam, M.K.; Vinsen, K.; Sostaric, T.; Lim, L.Y.; Locher, C. Detection of syrup adulterants in manuka and jarrah honey using HPTLC-multivariate data analysis. *PeerJ* **2021**, *9*, e12186. [CrossRef]
138. Luca, L.; Pauliuc, D.; Oroian, M. Honey microbiota, methods for determining the microbiological composition and the antimicrobial effect of honey—A review. *Food Chem. X* **2024**, *23*, 101524. [CrossRef] [PubMed]
139. Hungerford, N.L.; Carter, S.J.; Anuj, S.R.; Tan, B.L.L.; Hnatko, D.; Martin, C.L.; Sharma, E.; Yin, M.; Nguyen, T.T.P.; Melksham, K.J.; et al. Analysis of pyrrolizidine alkaloids in Queensland honey: Using low temperature chromatography to resolve stereoisomers and identify botanical sources by UHPLC-MS/MS. *Toxins* **2019**, *11*, 726. [CrossRef] [PubMed]
140. Xu, J.; Wang, W.; Yang, X.; Xiong, A.; Yang, L.; Wang, Z. Pyrrolizidine alkaloids: An update on their metabolism and hepatotoxicity mechanism. *Liver Res.* **2019**, *3*, 176–184. [CrossRef]

 https://doi.org/10.3390/biotech15010005

141. Hossain, M.L.; Hammer, K.; Lim, L.Y.; Hettiarachchi, D.; Locher, C. Optimization of an agar overlay assay for the assessment of the antimicrobial activity of topically applied semi-solid antiseptic products including honey-based formulations. *J. Microbiol. Methods* **2022**, *202*, 106596. [CrossRef]
142. Islam, M.K.; Sostaric, T.; Lim, L.Y.; Hammer, K.; Locher, C. Sugar profiling of honeys for authentication and detection of adulterants using high-performance thin layer chromatography. *Molecules* **2020**, *25*, 5289. [CrossRef]
143. Roshan, N.; Rippers, T.; Locher, C.; Hammer, K.A. Antibacterial activity and chemical characteristics of several Western Australian honeys compared to manuka honey and pasture honey. *Arch. Microbiol.* **2017**, *199*, 347–355. [CrossRef] [PubMed]
144. Brar, D.S.; Pant, K.; Krishnan, R.; Kaur, S.; Rasane, P.; Nanda, V.; Saxena, S.; Gautam, S. A comprehensive review on unethical honey: Validation by emerging techniques. *Food Control* **2023**, *145*, 109482. [CrossRef]
145. Liu, T.; Ming, K.; Wang, W.; Qiao, N.; Qiu, S.; Yi, S.; Huang, X.; Luo, L. Discrimination of honey and syrup-based adulteration by mineral element chemometrics profiling. *Food Chem.* **2021**, *343*, 128455. [CrossRef]
146. Ahmed, E. Detection of honey adulteration using machine learning. *PLoS Digit. Health* **2024**, *3*, e0000536. [CrossRef]
147. Fakhlaei, R.; Selamat, J.; Khatib, A.; Razis, A.F.A.; Sukor, R.; Ahmad, S.; Babadi, A.A. The Toxic Impact of Honey Adulteration: A Review. *Foods* **2020**, *9*, 1538. [CrossRef]

https://doi.org/10.3390/biotech15010005

MDPI AG
Grosspeteranlage 5
4052 Basel
Switzerland
Tel.: +41 61 683 77 34

BioTech Editorial Office
E-mail: biotech@mdpi.com
www.mdpi.com/journal/biotech